Mesay Solomon Tesema, Digafie Zeleke

# Practical Chemistry

## Also of Interest

*Industrial Inorganic Chemistry*
Mark Anthony Benvenuto, 2024
ISBN 978-3-11-132944-4, e-ISBN 978-3-11-132951-2

*Applied Inorganic Chemistry*
*Volume 1–3*
Rainer Pöttgen, Thomas Jüstel and Cristian A. Strassert (Eds.), 2022
Set-ISBN 978-3-11-074233-6

*Inorganic Chemistry.*
*Some New Facets*
Ram Charitra Maurya, 2021
ISBN 978-3-11-072725-8, e-ISBN 978-3-11-072728-9

*Rare Earth Chemistry*
Rainer Pöttgen, Thomas Jüstel and Cristian A. Strassert, 2020
ISBN 978-3-11-065360-1, e-ISBN 978-3-11-065492-9

Mesay Solomon Tesema, Digafie Zeleke

# Practical Chemistry

Transition Metals

Volume 2

DE GRUYTER

**Authors**
Mesay Solomon Tesema
Department of Chemistry
College of Natural Sciences
Salale University
P.O. Box 245
Oromia, Fiche
Ethiopia
mesaykiyya@gmail.com

Dr. Digafie Zeleke
Department of Chemistry
College of Natural Sciences
Salale University
P.O. Box 245
Oromia, Fiche
Ethiopia
digafiezel@gmail.com

ISBN 978-3-11-157384-7
e-ISBN (PDF) 978-3-11-157434-9
e-ISBN (EPUB) 978-3-11-157488-2

**Library of Congress Control Number: 2024944100**

**Bibliographic information published by the Deutsche Nationalbibliothek**
The Deutsche Nationalbibliothek lists this publication in the Deutsche Nationalbibliografie; detailed bibliographic data are available on the internet at http://dnb.dnb.de.


Cover image: angelp/iStock/Getty Images Plus
Typesetting: Integra Software Services Pvt. Ltd.

www.degruyter.com

# Preface

This manual has been prepared for *CHEM 3033 Inorganic Chemistry Laboratory Manual* and includes 12 experiments, which are related to the topics covered in the course "CHEM 3031 Inorganic Chemistry II." The main purpose of this laboratory is to provide the students an appreciation for the occurrence, preparation, reaction, and properties of transition metals. It is also aimed to provide the students a degree of competence in the laboratory skills required for accurate and precise chemical analysis. Therefore, it is expected that the students will demonstrate proficiency in the theory underlying analytical techniques and will apply this theory to obtain reliable analytical results on the preparation of some transition metals.

https://doi.org/10.1515/9783111574349-202

# Contents

# General laboratory safety

## Introduction

At all times, the instructor and students should observe safety rules. They should always wear safety glasses in the laboratory and should become familiar with emergency treatment. Laboratories are places of great responsibility. Careful practice and mature behavior can prevent most mishaps. The following are all very important. Treating the lab with respect makes it far less dangerous.

**Eye protection:** Goggles or safety glasses must be worn at all times. Eyeglasses with shatterproof glass are inadequate without goggles or safety glasses. Side shields are required for all protective eyewear.

**Shoes:** Shoes that completely cover the feet are required in the laboratory.

**Protective clothing:** A protective apron or lab coat is recommended in the laboratory. If any chemical is spilled on your skin or clothing, it must be washed off immediately.

**Food and drink:** Food and beverage are strictly prohibited in the laboratory. Do not taste or smell any chemical.

**No unauthorized experiments:** Do not perform any unauthorized experiments. Chemicals, supplies, or equipment must not be removed from the laboratory. All experiments must be approved by the instructor.

**Smoking:** Smoking is prohibited in the laboratory.

**Personal item:** No bags, coats, books (except the lab book), or laptop computers should be brought into the laboratory. Ask your instructor where these items can be stored while you are in the laboratory. Bring in only the items that are needed during the laboratory period. These items can be damaged by the chemicals in the laboratory.

**Use of equipment:** Do not use any equipment until the instructor has shown you how to use it.

**Glassware:** Do not use any broken, chipped, or cracked glassware. Get replacement glassware from your instructor.

**Bench cleanup:** At the end of the laboratory period, put away all equipment, clean the laboratory bench, and wash your hands.

**Use of chemicals:** Take only the amount that is needed. Leave all bottles in their proper places. Place the lids on the bottles after use. Clean up all spilled chemicals immediately.

**Careful reading of labels:** A material safety data sheet is available for each chemical in the laboratory. Ask your instructor where the paper copies are located. Material

https://doi.org/10.1515/9783111574349-204

safety data sheets are also available on the web. Many chemical companies have posted this information. Use web search engines to locate this information. Students are encouraged to obtain this information prior to using the chemical in the laboratory. Safety, health, and fire precautions are the most important information to locate. Special instructions for the handling of certain reagents may be posted by the instructor.

**Procedures for proper labeling**

- Manufacturer chemical labels must never be removed or defaced until the chemical is completely used.
- All chemical and waste containers must be clearly labeled with the full chemical name(s) (no abbreviations or formulas) and must contain appropriate hazard warning information.
- Small containers that are difficult to label such as 1–10 mL vials and test tubes can be numbered, lettered, or coded as long as an associated log is available that identifies the chemical constituents. Groups of small containers can be labeled as a group and stored together.
- Unattended beakers, flasks, and other laboratory equipment containing chemicals used during an experiment must be labeled with the full chemical name(s).
- All chemicals should be labeled with the "date received" and "date opened."
- All laboratory chemical waste containers must be labeled with the name of the chemicals contained.

**Waste disposal:** In recent years, the rules regarding waste disposal have become more rigidly defined. Reagents are never poured down the sink. Containers for chemical wastes are provided in the laboratory. Different containers are needed for different types of waste chemicals, such as chlorinated hydrocarbons, hazardous materials, and metals. All reagents in the waste container are listed on the container.

**Fume hoods:** Most laboratories provide fume hood areas or bench-top fume hoods. Always use these hoods. If you think the hoods are not turned on, then bring to the attention of your instructor. Often students are provided with simple methods of testing hood efficiency, and these should be used periodically. Safety regulations usually prohibit storage of toxic substances in hoods, and fume cupboards for such compounds are normally available.

**Gloves:** Most laboratories provide boxes of gloves. Modem gloves are quite manageable and allow for handling of equipment with some agility. Gloves have their place and can certainly protect your hands from obnoxious odors or chemicals that can cause allergic responses. But they are not a license for sloppy technique. Moreover, they are often easily penetrated by some compounds. Due care is still required.

**Compressed gas cylinders:** Compressed gas cylinders, especially those that are nearly as tall as an adult, can be dangerous if not clamped to the bench top. Gas cylinders containing inert gases such as nitrogen or helium may well be around the lab. Cylinders containing chlorine or more toxic reagents should be stored in a fume cupboard.

**Safety equipment:** The location of safety equipment should be made known to you. Moreover, you should know if and when you should use these. Most of the following items should be readily available in the chemistry laboratory, and items on this list or their description may vary due to local safety regulations:
- Fire blanket
- Fire extinguisher
- Eye-wash fountain
- Shower and first aid kit washes for acid or base (alkali) bums

**Accident reporting:** All accidents should be reported. The manner in which they should be reported will be provided by the instructor. It is also important that someone accompany an injured person who is sent out of the laboratory for special care; if the injured person should faint, the injury could easily become compounded. Medical treatment, except in the simplest of cases, is usually not the responsibility of the instructor. Very simple, superficial wounds can be cleaned and bandaged by the instructor. But any reasonably serious treatment is the job of a medical professional. The student should be sent to the college medical center accompanied by someone from the chemistry department. In all labs, the instructor should provide the students with instructions that are consistent with local regulations.

## Chemical hazard pictogram

Oxidizers

Flammables, Self Reactives, Pyrophorics, Self-Heating, Emits Flammable Gas, Organic Peroxides

Explosives, Self Reactives, Organic Peroxides

Acutely Toxic (severe)

Burns Skin, Damages Eyes, Corrosive to Metals

Gases Under Pressure

Carcinogen, Respiratory Sensitizer, Reproductive Toxicity, TargetOrgan Toxicity, Mutagenicity Aspiration Toxicity

Toxic to aquatic environment

Acutely toxic(harmful), Irritant to skin, eyes or respiratory tract, Skin sensitizer, Hazardous to the Ozone layer.

# Laboratory report writing format

**1. Cover page**

Name of university-----
College------------------
Department-------------
Practical course name---------
Group --------    subgroup-----
Name of members    Id no.

Submission date

**2. Body of the report**

- Experiment number: _______
- Title: _______
- Objective: (should be clear and short)
- Theory: (not more than half a page)
- Procedure
- List of chemicals and apparatus
- Observation: Compile the possible observation made (solubility, precipitates, color change, gas evolution, etc.)
- Results: (write the balanced equation of the chemical reaction in ionic form)
- Discussion: To discuss on outcomes the student refer the theoretical parts. Explain the errors if any.
- Conclusion: Write statements that explain your objective was fulfilled.
- References: Write down your sources except manual.

For example, you can write the observation and result parts as the following sample.

# The chemistry of titanium

1 Preparation and properties of titanic acid

1.1 Reaction of titanyl sulfate with aqueous NaOH and ammonia: preparation of orthotitanic acid

$$TiO_2{}^{+} + 2OH^{-} + H_2O \rightarrow TiO_2 \cdot 2H_2O$$

Observation: . . . . . . . . . . . . . . . . . . . . . . . . . . . .

Result: With both dilute aqueous NaOH and ammonia, white precipitate of $TiO_2 \cdot 2H_2O$, orthotitanic acid, is best described as hydrated $TiO_2$.

1.2 Reaction of titanyl sulfate with ammonium sulfate

$$S_2{}^{-} + H_2O \rightarrow HS^{-} + OH^{-} \text{(protolysis)}$$

https://doi.org/10.1515/9783111574349-205

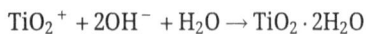

$$TiO_2{}^{+} + 2OH^{-} + H_2O \rightarrow TiO_2 \cdot 2H_2O$$

Observation. . . . . . . . . . . . . . . . . . . . . . . . . . . . . . . . .

Result: White precipitate of orthotitanic acid to protolysis of the sulfide.

1.3 Behavior of orthotitanic acid with respect to sulfuric acid and NaOH:

$$TiO_2{}^{+} \cdot 2H_2O + 2H^{+}$$
$$\rightarrow TiO_2{}^{+} + 3H_2O$$

Observation. . . . . . . . . . . . . . . . . . . . . . . . . . . . . .

Result: Freshly prepared orthotitanic acid is dissolved by dilute $H_2SO_4$ under the formation of colorless $TiO_2{}^{+}$. $TiO_2 \cdot 2H_2O$ is insoluble in dilute NaOH, although $TiO_2 \cdot 2H_2O$ has amphoteric properties that seem to predominate.

1.4 Conversion of orthotitanic acid into metatitanic acid

$$TiO_2 \cdot 2H_2O \xrightarrow{heat} TiO_2 \cdot 2H_2O + H_2O$$

Observation. . . . . . . . . . . . . . . . . . . . . . . . . . . . . . . . . . . . . . . . . .

Result: On boiling, orthotitanic acid ($TiO_2 \cdot 2H_2O$) gets converted into metatitanic acid ($TiO_2 \cdot 2H_2O$).

The latter is insoluble in dilute $H_2SO_4$ (difference to $TiO_2 \cdot 2H_2O$).

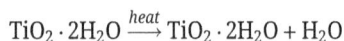

# 1 General aspects of the chemistry of transition elements

## 1.1 Position in the periodic table

The periodic table is an arrangement of the entire element in periods and groups according to their atomic number and electronic configuration. The latter follows naturally from the Aufbau principle. An important generalization concerning the valence electronic configuration is that the free atoms of a group in the periodic chart have the same number of electrons in their outermost shell (valence electron) and usually also have the same electronic configuration.

Therefore, knowledge of the position of an element in the periodic table (rather than its atomic number) is for a chemist as essential as the multiplication table for a mathematician. Depending on the kind of subshell being occupied by the valence electrons, the elements may be classified into three major groups (see figure 1.1):

1. Main group elements (s- and p-block elements) – within the outermost shell only s- and p-subshells are populated: $ns^a np^b (a = 1, 2; b = 0 - 6)$
2. Transition elements (d-block elements) – the orbital of the last but one principal shell are occupied: $(n-1)\mathrm{d}^c ns^2 (c = 0 - 10)$
3. Lanthanides and actinides; inner transition elements (f-block elements) – The f-levels of the outermost but two principal shells are filled: $(n-2)\mathrm{f}^d (n-1)\mathrm{d}^1 ns^2 (d = 1 - 14)$

Of course, the main group elements can be further subdivided into s-block (all of them being light metals) and p-block. The latter is composed of the so-called post-transition metals and nonmetals.

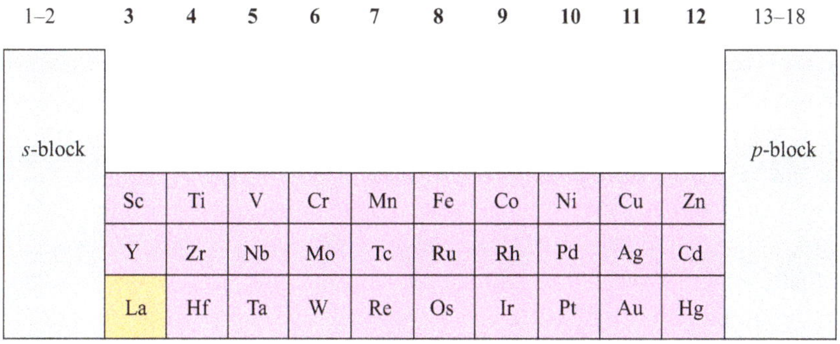


Figure 1.1: General sketch of the position of transition metals in the periodic table.

The electronic configuration is the distribution of electrons in the shell. Transition metals have the scheme that ends with $(n-1)\mathrm{d}^c ns^2 (c = 0 - 10)$. The division of the first

https://doi.org/10.1515/9783111574349-001

transition series into two "sets" is clearly related to the filling of the d-orbitals – at the dividing element, manganese, the 3d level is half-filled (one electron in each d orbital). Thereafter the singly occupied d orbitals become double-filled until filling is complete at copper and zinc. The fact that the configurations $3d^5$ (half-full) and $3d^{10}$ (full) are obtained at chromium and copper, respectively, in each case (see Table 1.1) (at the cost of removing an electron from the 4s level) suggests that these configurations $3d^5$ and $3d^{10}$ are particularly "stable."

Table 1.1: Electronic configuration of the transition elements.

| **Group** | **3** | **4** | **5** | **6** | **7** | **8** | **9** | **10** | **11** | **12** |
|---|---|---|---|---|---|---|---|---|---|---|
| **Element** | Sc | Ti | V | Cr | Mn | Fe | Co | Ni | Cu | Zn |
| **Configuration** | $3d^1 4s^2$ | $3d^2 4s^2$ | $3d^3 4s^2$ | $3d^5 4s^1$ | $3d^5 4s^2$ | $3d^6 4s^2$ | $3d^7 4s^2$ | $3d^8 4s^2$ | $3d^{10} 4s^1$ | $3d^{10} 4s^2$ |
| **Element** | Y | Zr | Nb | Mo | Tc | Ru | Rh | Pd | Ag | Cd |
| **Configuration** | $4d^1 5s^2$ | $4d^2 5s^2$ | $4d^4 5s^1$ | $4d^5 5s^1$ | $4d^5 5s^2$ | $4d^7 5s^1$ | $4d^8 5s^1$ | $4d^{10} 5s^0$ | $4d^{10} 5s^1$ | $4d^{10} 5s^2$ |
| **Element** | Lu | Hf | Ta | W | Re | Os | Ir | Pt | Au | Hg |
| **Configuration** | $5d^1 6s^2$ | $5d^2 6s^2$ | $5d^3 6s^2$ | $5d^4 6s^2$ | $5d^5 6s^2$ | $5d^6 6s^2$ | $5d^7 6s^2$ | $5d^9 6s^1$ | $5d^{10} 6s^1$ | $5d^{10} 6s^2$ |

It is mainly due to the strong interelectronic interaction in the more compact 3d orbitals that the energetically less stable 4s orbital remains doubly occupied in Sc, Ti, and V. In fact, the 3d orbitals are significantly more stabilized with increasing nuclear charge across the period than the 4s-level. As a consequence, in the case of chromium, one electron from 4s orbital drops into the 3d-level giving rise to a $3d^5\ 4s^1$ configuration. Because of the relatively high spin pairing energy in the compact 3d-orbitals (compared to the more extended 4s-orbital), the next electron prefers again to stay in the energetically higher 4s level. Therefore, manganese exhibits the configuration $3d^5\ 4s^2$. Arriving at copper, a similar anomaly is to be observed. One 4s-electron has fallen into the now considerably more stable 3d-level despite interelectronic repulsion and spin pairing. The situation becomes even more complex for the 4d and 5d elements.

Three more characteristics shall be mentioned here:

All transition metal ions involved in chemical bonding possess a $d^n$ configuration, for example, $Ti^{3+}$ in $[Ti(H_2O)_6]^{3+}$ is a $3d^1$ rather than a $4s^1$ system. This can be attributed to the interaction between the central ion and ligands causing a general destabilization of the $(n–1)$d and $n$s orbitals, which is, however, greater for large $n$s than for the more compressed $(n–1)$d orbitals. Half and fully occupied subshells represent relatively stable energy states of the electronic system. Recall that spin pairing is an endothermic process. Thus, the addition of a sixth electron to the d5 system requires some extra energy – the so-called spin pairing energy.

For example, $MnO_4$ in acidic medium is normally reduced to $Mn^{2+}$ ($d^5$) rather than $Mn^6$ ($d^6$), or that $Fe^{2+}$ ($d^6$) is readily oxidized to $Fe^{3+}$ ($d^5$) under various conditions. In case of a fully occupied subshell, it is clear that adding another electron means populating an energetically higher sub- or principal level.

Due to lanthanide contraction, the 5d elements following lutetium have drastically reduced the atomic and ionic radii, being quite close to the radii of the corresponding 4d elements. due to the same chemical and physical properties elements are couples together. for example Y-La, Zr-Hf, Nb-Ta, and Mo-W and, to a smaller extent, also those of the following 4d–5d element pairs turn out to be rather similar but differ markedly from the behavior of the 3d elements.

### 1.1.1 Oxidation states

As a consequence of their electronic configuration, compounds of transition elements exist in various oxidation states. Thus, stable manganese compounds have been isolated with Mn in all oxidation states from **–III to +VII**. Table 1.2 summarizes the most common and some less frequently encountered positive oxidation numbers (ONs) of the transition metals.

There are two main trends to be emphasized:

- The stability of the high ONs increases down the group.
- The stability of the highest possible ON decreases across a period from left to right.

For the elements Ti, V, Cr, and Mn, it is possible to realize the maximum ON in aqueous solution ($TiO^{2+}$, $VO_4^{3-}$, and $MnO^{4-}$). However, in the case of iron and the following 3d elements, the ionization energy required to generate the highest oxidation state is increasing rapidly and can no longer be compensated by exothermic processes such as hydration, bond formation, and lattice energy.

Table 1.2: Electronic configuration, oxidation number, and color of some transition metal ions.

| 3d transition metal ions | Electronic configuration | Oxidation number | Color |
|---|---|---|---|
| $TiO^{2+}$ | $d^0$ | +IV | Colorless |
| $VO4^{3-}$ | $d^0$ | V | Colorless |
| $CrO_4^{2-}$ | $d^0$ | +VI | Yellow |
| $VO_3^-$, $VO_2^-$ | $d^0$ | V | Colorless to yellow |
| $Cr_2O_7^{2-}$ | $d^0$ | +VI | Orange |
| $Ti^{3+}$ | $d^1$ | III | Violet |
| $VO^{2+}$ | $d^1$ | +IV | Blue |
| $V^{3+}$ | $d^2$ | +III | Green |
| $V^{2+}$ | $d^3$ | +II | Violet |
| $Cr^{3+}$ | $d^3$ | +III | Violet (green) |

Table 1.2 (continued)

| 3d transition metal ions | Electronic configuration | Oxidation number | Color |
|---|---|---|---|
| $MnO_2$ | $d^3$ | +IV | Brown solid |
| $Cr^{2+}$ | $d^4$ | +II | Blue |
| $Mn^{3+}$ | $d^4$ | +III | Violet |
| $Mn^{2+}$ | $d^5$ | +II | Pink |
| $Fe^{3+}$ | $d^5$ | +III | Yellow |
| $Fe^{2+}$ | $d^6$ | +II | Green |
| $Co^{3+}$ | $d^6$ | +III | Red |
| $Co^{2+}$ | $d^7$ | +II | Pink |
| $Ni^{2+}$ | $d^8$ | +II | Green |
| $Cu^{2+}$ | $d^9$ | +II | Blue |
| $Cu^{+}$ | $d^{10}$ | +I | Colorless |
| $Zn^{2+}$ | $d^{10}$ | +II | Colorless |

In Table 1.2:
- The colors of $M^{2+}$ are those of hexaquo derivatives.
- $MnO_2$ is included as it is frequently encountered in redox reactions and is the only Mn(IV) compound to be dealt with in the experimental part.

### 1.1.2 Color

Transition metal compounds are very often colored; frequently (but not always) the color is due to the presence of coordination complexes. When a cation containing d electrons is surrounded by other ions or polar molecules, either in a complex ion in solution or in a solid, a splitting of the energy levels of the five d-orbitals (all originally having the same energy) occurs. When light falls on such a system, electrons can move between these split levels. The energy absorbed in this process corresponds to the absorption of the light at certain wavelengths, usually in the visible part of the spectrum; hence, color is observed. For a given cation, the kind of absorption produced its intensity, and the position in the spectrum depends very much on the coordination number and the surrounding ligands. We can illustrate this with reference to the $Cu^{2+}$ ion. In solid anhydrous copper(II) sulfate, the $Cu^{2+}$ ion is surrounded by $SO_4^{2-}$ ions. In this environment, the d-orbital splitting is such that the absorption of light by the $Cu^{2+}$ cation is not in the visible part of the spectrum, and the substance appears white. If the solid is now dissolved in water, the $Cu^{2+}$ ion becomes surrounded by water molecules, and complex species such as $Cu(H_2O)^{6+}$ are formed. These absorb light in the visible part of the spectrum and appear pale blue. If this solution of copper(II) sulfate is allowed to crystallize, water molecules remain coordinated around the $Cu^{2+}$ ion in the solid copper(II) sulfate pentahydrate ($CuSO_4{\cdot}5H_2O$) and the solid is pale blue. When an excess of ammonia is added to the original solution, some of the water ligands around the copper(II) ion are replaced by ammonia:

$$[Cu(H_2O)_6]^{2+}\text{(pale blue)} + 4NH_3 \longrightarrow [Cu(NH_3)_4(H_2O)_2]^{2+}\text{(deep blue)} + 4H_2O$$

When ligands surround a metal ion, they split its d-orbital energies into two levels. The color we see comes from the light not absorbed, which depends on the metal, ligands, and complex geometry. Strong-field ligands often result in blue or green colors, while weak-field ligands can lead to red or orange hues. (In transition metals, the d-orbitals are essential because they are commonly utilized in bonding.) If excess chloride ion is added to a blue solution containing $[Cu(H_20)_6]^{2+}$, he-re the new splitting results in a yellow-green color:

$$\text{th}[Cu(H_2O)_6]^{2+} + 4Cl^- \rightarrow [CuCl_4]^{2-} + 6H_2O$$

Distorted Distorted

Octahedral Tetrahedral

Pale blue Yellow

The d-orbital splitting depends on the oxidation state of a given ion; hence, two complex ions with the same shape, ligands, and coordination number can differ in color.

### 1.1.3 Coordination complexes

Coordination compound is a compound formed when a central metal atom or ion is surrounded (coordinated) by a number of anions or molecules in such a way that the number of the coordinated anions or molecules exceeds the normal covalence of the central atom or ion. It will be recalled that they are defined (and named) in terms of (a) the central metal atom or ion and its oxidation state, (b) the number of surrounding ligands which may be ions, atoms, or polar molecules, and (c) the overall charge on the complex, determined by the oxidation state of the central atom and the charges (if any) on the ligands. Some examples are (Table 1.3):

Table 1.3: Oxidation state of central metal.

| Oxidation state of central metal ion | Examples | Names |
|---|---|---|
| 7 | $Mn0_4^-$ | Permanganate, but better manganite(VII) (strictly, tetraoxomanganate(VII)) |
| 6 | $CrO_4^{2-}$ | Chromate, better chromate(VI) (strictly, tetraoxochromate (VI)) |
| 4 | $TiCl_4$ | Titanium tetrachloride or tetrachlorotitanium(IV) |
| 3 | $[Fe(CN)_6]^{3-}$ | Hexacyanoferrate(III) |
| 2 | $[Ni(NH_3)_6]^{2+}$ | Hexaamminenickel(II) |
| 0 | $Fe(CO)_5$ | Iron pentacarbonyl or pentacarbonyl iron(0) |

**Note:** Those complexes can have negative, positive, or zero overall charge. $MnO_4^-$ and $CrO_4^-$ are usually considered to be oxo acid anions but there is no essential difference between these and other complexes. For example, the anion $MnO_4^-$ can be regarded formally as a manganese ion in oxidation state +7 surrounded by four oxide ion ($O^{2-}$) ligands (in fact of course there is covalent bonding between the oxide ligands and the $Mn^{VII}$ ion, leading to partial transfer of the negative charges of oxide to the manganese). In general, high oxidation states (e.g., those of manganese +7 and chromium +6) are only found in oxides (e.g., $Mn_2O_7$ and $CrO_3$), oxo acid anions $(MnO_4^-, CrO_4^{2-}, Cr_2O_7^{2-})$, and sometimes fluorides (there is no $MnF_7$ known, but $CrF_6$ is known). Hence, the number of complexes in high oxidation states is very limited. At lower oxidation states, a variety of ligands can form complexes, and some common ligands are:

$$H_2O \text{ in } [Fe(H_2O)_6]^{2+},\ NH_3 \text{ in } [Co(NH_3)_6]^{3+},\ CN^- \text{ in } [Ni(CN)_4]^{2-}, Cl^- \text{ in } [CuCl_4]^{2-}$$

However, stable complexes where the oxidation state of the central metal atom is 0 are only formed with a very few ligands, notably carbon monoxide (e.g., $Ni(CO)_4$ and $Fe(CO)_5$) and phosphorus trifluoride, $PF_3$ (e.g., $Ni(PF_3)_4$).

### 1.1.4 Interstitial compounds

The transition metal structures consist of close-packed arrays of relatively large atoms. Between these atoms, in the "holes," small atoms, notably those of hydrogen, nitrogen and carbon, can be inserted, without very much distortion of the original metal structure, to give interstitial compounds (e.g., hydrides). As the metal structure is "locked" by these atoms, the resulting compound is often much harder than the original metal, and some of the compounds are therefore of industrial importance. Since there is a definite ratio of holes to atoms, filling of all the holes yields compounds with definite small atom–metal atom ratios; in practice, all the holes are not always filled, and compounds of less definite composition (nonstoichiometric compounds) are formed.

### 1.1.5 Acid–base reactions

A simple transition metal salt with $x^-$ being the anion of strong acid, for example, $Cl^-, NO_3^-$, or $\frac{1}{2} SO_4^{2-}$, undergoes protolysis when dissolved in water. The solution of these salts behaves as slightly acidic, due to the partial dissociation of the aqua complexes into hydroxo species and $H_3O^+$:

$$[M(H_2O)_6]^{n+} + H_2O \rightleftharpoons [M(H_2O)_5(OH)]^{(n-1)+} + H_3O^+$$

The extent of protolysis increases with the oxidation state of $M^{n+}$. Accordingly simple aqua cations with the central in a high positive ON are in general unstable with respect to protolysis in aqueous solution and are immediately converted into the more stable oxo cations or oxo anions depending on the pH value of the system. Hydrolysis may also result in the precipitation of insoluble hydroxides or oxides. As an example, the hexaquotitanium(IV) cation does not exist in the aqueous solutions of Ti(IV) salts. Depending on the pH vale, hydroxo complexes or the hydrated $TiO^{2+}$ ion can be detected:

$$[Ti(H_2O)_6]^{4+} \underset{-2H_3O+}{\overset{+2H_2O}{\rightleftharpoons}} [Ti(H_2O)_4(OH)_2]^{2+} \rightleftharpoons TiO^{2+}(aq)$$

Other examples for this phenomenon are the various ions of chromium in the oxidation states +II, +III, and +VI. The hexaquo complexes of $Cr^{2+}$ and $Cr^{3+}$ are stable and only slightly protolyzed.

However, in the oxidation state +VI, it is the oxo anions $CrO_4^{2-}$ and $Cr_2O_7^{2-}$ that are exclusively encountered:

$$\left[\overset{-e^-}{Cr}(H_2O)_6\right]^{2+} \longrightarrow [Cr(H_2O)_6]^{3+} \xrightarrow{-3e^-} CrO_4^{2-} + 4H_2O + Cr_2O_7^{2-}$$

Of great significance for transition metal is the behavior of its oxide with its acids and bases:

$$M_2O_3 + 6H^+ \longrightarrow 2M^{3+} + 3H_2O$$

$$M_2O_3 + 6OH^- \longrightarrow 2MO_3^{3-} + 2H_2O$$

In the first equation, the oxide is said to be a base anhydride in contrast to the second one, where it behaves like an acid anhydride. Many transition metal oxides display amphoteric properties. In principle, the basic character of a transition element increases with decreasing ON. Thus, V(II) and V(III) form simple vanadium salts in acidic solution. In alkali medium, the sparingly soluble hydroxides that do not dissolve in an excess of base are precipitated because there is no tendency to form oxo or hydroxo vanadates. A similar behavior is found for most of the transition metal $M^{2+}$ and $M^{3+}$ ions. Now, consider vanadium in its highest oxidation state. The oxide $V_2O_5$ is amphoteric and reacts in acidic solution to give $VO_2^+$, whereas in basic medium orthovanadates(V) are obtained:

$$V_2O_5 + 2H^+ \longrightarrow 2VO_2^+ + H_2O$$

$$V_2O_5 + 6OH^+ \longrightarrow 2VO_4^{3-} + 3H_2O$$

Another interesting phenomenon encountered with some transition elements, especially those of the VB and VIB group, is the tendency to undergo various condensation reactions, yielding the so-called isopolyanions or isopolyacids. Some such interconversions are depicted in the following equations:

$$2VO_4^{3-} + 2H^+ \rightleftharpoons V_2O_7^{4-} + H_2O \qquad \text{(divanadate)}$$

$$2VO^{4-}{}_7 + 4H^+ \rightleftharpoons V_4O_{12}{}^{4-} + 2H_2O \qquad \text{(tetravanadate)}$$

$$5V_4O_{12}{}^4 + 8H^+ \rightleftharpoons 2VO^{6-} + 4HO \qquad \text{(decavanadate)}$$

$$2VO^{6-} + 6H^+ \rightleftharpoons 5VO(aq) \qquad \text{(hydrous vanadium(V)oxide)}$$

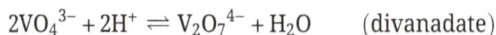

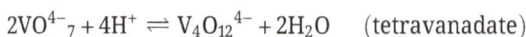

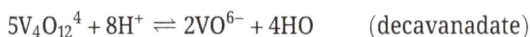

With decreasing pH value of the solution, the degree of condensation is enhanced. The process can be monitored nicely by the characteristic color change from colorless ($VO_4^{3-}$), via yellow ($V_2O_7^{4}$), and orange (polyvanadates), to orange colored (polymeric $V_2O_5$). In highly acidic medium, the oxide dissolves again, yielding the yellow vanadyl cation, $VO_2^+$.

### 1.1.6 Redox reactions

The great variety of oxidation states in which transition metals may exist give rise to numerous redox reactions. It has very often been observed that students are terrified by balancing such equations along with aqueous solutions, and these are most appropriately written as ionic equations. Hence, basic rules and procedures are summarized in this section.

In tackling the problem, three questions have to be considered:

- How does one determine ONs?
- When does one write ions and ionic equations?
- How are redox equations completed and bleached?

### 1.1.7 Oxidation number

The ON of an element in a molecule or molecular ion is estimated by splitting the latter hydrolytically into atomic ions according to their relative electronegativities. The ON is indicated on the top of the element symbol which is a plus or minus sign followed by the number of charge (Roman or Arabic symbol). Notice that ionic charges are written in a similar manner but with the number at followed by the ± sign. In order to come up with correct result, stick to the rules below:

1. In element compounds, the ON is always zero (e.g., graphite, $O_2$, $O_3$, $H_2$, $S_8$, and metals).
2. The ON of fluorine in compounds must be –1 because it is the most electronegative element of all.
3. Hydrogen usually gets the ON +1. An exception is hydrides such as NaH or $LiAlH_4$, where hydrogen is attached to highly electronegative elements. Hydride hydrogen has the ON –1.

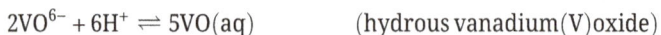

4. The ON –2 is assigned to oxygen in most of its compounds. The only exception is peroxo groups with oxygen possessing in ON –1; for example, in $H_2O_2$ and $Na_2O_2$, and certain fluoro compounds with +1 or +2 ONs for oxygen.
5. In element compounds, the ON is always zero (e.g., graphite, $O_2$, $O_3$, $H_2$, $S_8$, and metals).
6. The ON of fluorine in compounds must be –1 because it is the most electronegative element of all.
7. Hydrogen usually gets the ON +1. An exception is hydrides such as NaH or $LiAlH_4$, where hydrogen is attached to highly electronegative elements. Hydride hydrogen has the ON –1.
8. The ON –2 is assigned to oxygen in most of its compounds. The only exception is peroxo groups with oxygen possessing in ON –1, for example, in $H_2O_2$ and $Na_2O_2$, and certain fluoro compounds with +1 or +2 ONs for oxygen.
9. The remaining elements are also treated according to their electronegativity, taking into account that the sum of all ONs must be equal to zero for neutral molecules and equal to the ionic charge for ions.

The following examples illustrate the above rules:

$$\overset{+1}{Li}\,\overset{-1}{H},\ \overset{+2}{O}\,\overset{-1}{F_2},\ \overset{+1}{H_2}\,\overset{-1}{O},\ \overset{-1}{N}\,\underset{2}{\overset{+1}{H}} - \overset{-2}{O}\,\overset{+1}{H},\ \overset{+7}{Mn}\,\overset{-2}{O_4}{}_{-},\ \overset{+1}{K_2}\,\overset{+6}{Cr_2}\,\overset{-2}{O_7},\ \overset{-1}{CH_3}\,\overset{+1}{CH}\,\overset{-2}{O}$$

$S_2O_8{}^{2-}$ =

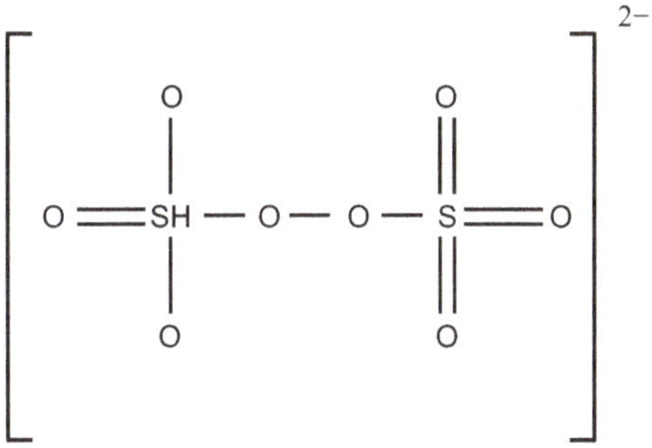

The peroxodisulfate anion contains sulfur in ON +6. Six O's are in the ON –2 but the remaining two O's of the peroxo bridge are in the oxidation state –1 (see rule 4).

## 1.1.8 Ionic equations

If ions are involved in a chemical reaction, ionic equations are the most appropriate to describe the process taking place. Moreover, ionic equations have the following advantages:

- They take into account that the chemistry in aqueous solution is mainly the chemistry of reacting ions.
- Omit all those particles that do not directly take part in the actual reaction.
- The stoichiometry and the balancing of ionic equations are always simpler than for overall equations.

In order to establish an ionic equation one has to distinguish the compounds dissociating completely in aqueous solutions and those that do not split into ions. Complete dissociation is found for:
- strong acids and bases, and
- salts.

Overall formulae are necessary for:
- Solids and precipitates that are insoluble in water, for example, $BaSO_4$, $Hg_2Cl_2$, $Fe(OH)_3$, sparingly soluble oxides and sulfides of transition metals, such as $MnO_2$, $TiO_2$, FeS, and $Ag_2O$.
- Gases that are merely physically dissolved in water or evolved in a reaction, such as $O_2$, $N_2$, $H_2$, NO, $NO_2$, and $Cl_2$; notice that HCl as a gas is a molecular compound that, however, dissociates completely into $H^+$ and $Cl^-$ after dissolution in water.
- Neutral and ionic complexes, for example, $[Fe\ (CN)_6]^{3/4-}$, $[Mn(H_2O)_6]^{2+}$, $[Ni(NH_3)_6]^{2+}$, and oxo anions such as $TiO^{2+}$, $VO_4^{3-}$, and $SO_4^{2-}$.
- Typical covalent compounds, for example, $CH_3OH$, $NH_2OH$, $NH_3$, and $H_2O$.
- In certain cases, one has to decide by means of the pH value of the solution that either the molecular or ionic formula has to be employed.
- Weak bases, for example, $NH_3$, are protonated in acidic medium (here formation of $NH_4^+$).
- Weak acids are gradually deprotonated in neutral and basic solution, for example, $H_2SO_3$ ($HSO_3^-$ and $SO_3^{2-}$) or $H_2S$ ($HS^-$ and $S^{2-}$). Other frequently encountered weak acids are $HNO_2$, HCN, $H_3PO_4$, carbonic acid, and acetic acid.

### 1.1.9 Redox equations

In a redox process, the ON of one element involved increases (loss of electron; oxidation), whereas the ON of another participant must be decreased (gain of electrons; reduction). Both half reactions are coupled with each other because the number of electrons lost in the oxidation has to be equal to the number of electrons consumed in the reduction. No electrons are left over.

In establishing a redox equation, the following operation should be strictly followed:
1. Selection of starting material and final products.
2. Determination of the ONs of all elements and selection of those that appear in different ONs on the left- and right-hand sides of the equation. These elements are oxidized and reduced, respectively. The ions and the molecules actually reacting, as established in this way, are of significance for the next step of the procedure. Set up the partial equations (half equations) for oxidation and reduction in a completely balanced manner.
3. Count the electron released and consumed, and find the common factor that balances the electrons.

4. Sum up both parallel equations and take care that the electrons are now canceled out completely.
5. Check the correctness of your equation:
   a. The sum of ionic charges must be equal on both sides.
   b. Count the element on both sides.
6. Balance the remaining elements beginning with oxygen followed by hydrogen.

Note that mixing the order of operations and mistakes in counting the electrons are the main reason for failure.

## 1.2 Worked out examples

1. Dissolution of titanium in dilute sulfuric acid under formation of the violet $Ti^{3+}$ cation $Ti + H^+ \longrightarrow Ti^{3+} + H_2$ starting and final products
   - Balancing the partial equations and finding the least common factor

$$Ti \longrightarrow Ti^{3+} + 3e^- \; x_2$$

$$2H^+ + 2e^- \longrightarrow H_2 \qquad x_3$$

   - Summing up the completed partial equations and final check

$$2Ti + 6H^+ \longrightarrow 2Ti^{3+} + 3H_2$$

Abbreviated procedure:

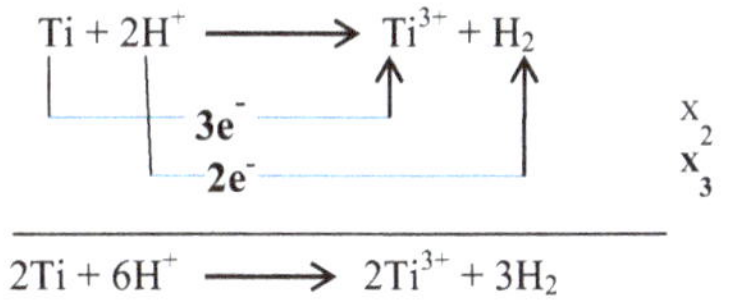


2. **Oxidation of $Fe^{2+}$ to $Fe^{3+}$ by $MnO_4^-$ in acidic solution**

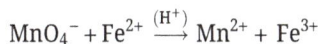

$$MnO_4^- + Fe^{2+} \xrightarrow{(H^+)} Mn^{2+} + Fe^{3+}$$

$$Fe^{2+} \longrightarrow Fe^{3+} + e^- \, x5$$

$$MnO^- + 8H^+ + 5e^- \longrightarrow Mn^{2+} + 4H_2O \qquad x1$$

$$5Fe^{2+} + MnO_4^- + 8H^+ \longrightarrow 5Fe^{3+} + Mn^{2+} + 4H_2O$$

In order to balance oxygen and hydrogen, pay attention to the following:
- If O and H are not directly involved in redox process their ON remains unchanged (i.e., −2 for O and + for H).

- In acidic medium, $H^+$ and $H_2O$ can be used for balancing, whereas in alkaline solution, only $H_2O$ and $OH^-$ are used.
- The oxide ion $O^{2-}$ itself is unstable in aqueous solution as it would immediately be converted into $H_2O$ (in acidic medium) or $OH^-$ (in basic medium):

$$O^{2-} + 2H^+ \longrightarrow H_2O, O^{2-} + H_2O \longrightarrow 2OH^-$$

3. **Oxidation of sulfurous acid by $MnO_4^-$ in acidic solution**
   $H_2SO_3$ is extremely unstable and in acidic solution it exists as $SO_2$ or $HSO_3^-$. In alkaline medium, however, the sulfite anion $SO_3^{2-}$ has to be employed:

$$SO_2 + H_2O \rightleftharpoons HSO_3^- + H^+ \; HSO_3^- + OH^- \rightleftharpoons SO_3^{2-} + H_2O$$

$$MnO_4 + HSO_3 \xrightarrow{(H^+)} Mn^{2+} + SO_4^{2-}$$

$$MnO^{4-} + 8H^+ + 5e^- \longrightarrow Mn^{2+} + 4H_2O \quad x2$$

$$HSO_3^- + H_2O \longrightarrow SO_4^{2-} + 3H^+ \quad x5$$

$$2MnO_4^- + 16H^+ + 5HSO_3^- + 5H_2O \longrightarrow 2Mn^{2+} + 8H_2O + 5SO_4^{2-} + 15H^+$$

After canceling $H_2O$ and $H^+$, the following equation is obtained:

$$2MnO_4^- + H^+ + 5HSO_3^- \longrightarrow 2Mn^{2+} + 3H_2O + 5SO_4^{2-}$$

### 1.2.1 Reduction of $MnO_4^-$ by $H_2O_2$ in acidic solution

$H_2O_2$ is known to react in three different ways:

$$H_2O_2 + 2e^- \longrightarrow 2OH^- \quad \text{(oxidizing agent)}$$
$$H_2O_2 \longrightarrow O_2 + 2H^+ + 2e^- \quad \text{(reducing agent)}$$
$$H_2O_2 \longrightarrow H_2O + \tfrac{1}{2}O_2 \quad \text{(disproportionation)}$$

$$MnO_4 + H_2O_2 \xrightarrow{(H^+)} Mn^{2+} + O_2$$

$$MnO_4^- + 8H^+ + 5e^- \longrightarrow Mn^{2+} + 4H_2O \quad x2$$

$$H_2O_2 \longrightarrow O_2 + 2H^+ + 2e^- \quad x5$$

$$2MnO_4^- + 5H_2O_2 + 6H^+ \longrightarrow 2Mn^{2+} + 5O_2 + 8H_2O$$

### 1.2.2 Oxidation of $Mn^{2+}$ by $MnO_4^-$ in alkaline solution (synproportionation)

In acidic medium, $MnO_4^-$ is reduced down to $Mn^{2+}$. Alkaline solutions stop the reduction, however, at Mn(IV). The only product possible under these conditions proves to black insoluble $MnO_2$:

$$Mn^{2+} + MnO_4^- \xrightarrow{(OH^-)} MnO_2$$

$$Mn^{2+} + 4OH^- \longrightarrow MnO_2 + 2H_2O + 2e^- \quad x3$$

$$MnO_4^- + 2H_2O + 3e^- \longrightarrow MnO_2 + 4OH^- \quad x2$$

$$3Mn^{2+} + 2MnO_4^- + 4OH^- \longrightarrow 5MnO_2 + 2H_2O$$

### 1.2.3 Reduction of $MnO_4^-$ by ethanol in alkaline solution

The determination of ONs in organic compound is necessary in order to count the electrons lost gain in a redox process. Though it is merely a formal matter without any chemical meaning, the procedure is the same as for ionic compounds:

$$C_2^{3+} O_4^{2-} \longrightarrow (C_2O_4 + 2e^-) \longrightarrow 2CO_2^{4+} + 2e^-$$

$$CH_3^{2-} - CH_2^{2-} - OH \longrightarrow CH_3 - CHO + 2H \longrightarrow CH_3^{1-} - CHO^{1-} + 2H + 2e^-$$

$$CH_3^{1-} - CHO^{1-} + H_2O^{2-} \longrightarrow CH_3 - COOH + 2H \longrightarrow CH_3^{0} - COOH^{0} + 2H + 2e^-$$

There are similar methods of determining ON yielding different absolute values. Thus, it is possible either to substitute the nonreacting alkyl chain by hydrogen or to neglect it completely. Notice that the same procedure is used for reactants and products. Independent of the method and the varying absolute ONs, the difference in the ONs, that is, the number of electrons exchanged in the oxidation or reduction process, is always the same:

$$MnO_4^- + C_2H_5OH \xrightarrow{(OH)} MnO_2 + CH_3CHO$$

$$MnO_4^- + 2H_2O + 3e^- \longrightarrow MnO_2 + 4OH^- \quad x2$$

$$C_2H_5OH + 2OH^- \longrightarrow CH_3CHO + 2H_2O + 2e^- \quad x3$$

$$2MnO_4^- + 3C_2H_5OH \longrightarrow 2MnO_2 + 3CH_3CHO + 2OH^- + 2H_2O$$

# Experiment 1

## 1.1 Chemistry of titanium

## 1.2 Objective

Students should be able to:
- observe characteristic of $Ti^{3+}$,
- observe hydrolysis of titanium(III) and titanium(IV),
- observe reducing properties of titanium(III) hydroxide, and
- discuss two methods for the preparation of titanium metal from $TiO_2$.

## 1.3 Theory

### 1.3.1 Occurrence

The Earth's crust contains approx. 0.6% Ti, which is therefore not a rare element. It occupies the ninth position in the frequency chart of elements and is 5–10 times more frequent than Cl, S, and P. As its ionic radius proves to be very similar to ions such as $Al^{3+}$, $Fe^{3+}$, and $Mg^{2+}$, respectively, Ti is found in minor amounts in almost each mineral, stone, or soil.

Typical titanium minerals are rutile, anatase, and brookite (three modification of $TiO_2$), as well as ilmenite ($FeTiO_3$) and perovskite ($CaTiO_3$). Less significant deposits are ilmenite-containing sands at sea coasts and $TiO_2$ as a constituent of clay. Moon rocks contain more Ti than earthly minerals.

In nature, titanium occurs only in the form of Ti(IV) compounds. However, lower ON of the elements such as +3, +2, and even 0 can be obtained synthetically.

Titanium has isotopes such as $^{46}Ti$ (7.9%), $^{47}Ti$ (7.3%), $^{48}Ti$ (73.9%), $^{49}Ti$ (5.5%), and $^{50}Ti$ (5.3%).

### 1.3.2 Preparation

Unlike the majority of other transition metals, titanium cannot be made by reduction of the oxide with carbon because of carbide formation. The reduction of $TiO_2$ by base metals, for example, Na, Ca, Mg, or hydrogen, does also not give the pure metal but yields lower titanium oxides. That is why most of the procedures go via titanium tetrahalides. $TiCl_4$ is prepared from oxide by reductive chlorination:

https://doi.org/10.1515/9783111574349-002

$$TiO_2 + 2C + 4Cl_2 \longrightarrow TiCl_4 + 2COCl_2$$

The following two equations show some laboratory methods for preparing (extracting) titanium:

$$TiO_2 + 2CaH_2 \longrightarrow Ti + 2CaO + H_2$$

$$TiCl_4 + 4Na \xrightarrow{800\,°C} Ti + 4NaCl$$

$$TiCl_4 + 2H_2 \xrightarrow{2{,}000\,°C} Ti + 4HCl$$

Large-scale industrial production of titanium is mainly done by the **KROLL** procedure:

$$TiCl_4 + Mg \xrightarrow{1{,}200\,°C} Ti + 2MgCl_2$$

High-purity titanium is obtained by the reaction of the crude metal with iodine and subsequent decomposition of the $TiI_4$ vapor at a hot filament according to the **Van Arkel** and **De Boer** process.

#### 1.3.2.1 Physical properties

Metallic titanium occurs in two allotropic (a structurally different form of an element) forms. Up to 882 °C, the α-modification is stable (hexagonal close-packed structure). At higher temperature, the latter is converted into the β-form (cubic body-centered structure). Ore titanium is ductile, silvery-white metal. The mechanical properties have a subtle dependence upon impurities. Thus, the Brinell hardness of pure titanium is twice that of the Ti containing only 0.003% nitrogen. Titanium is hardly compressible, has a low density of 4.4–4.5 g/cm, but a high mechanical strength. In particular, its density is lower than that of iron, its strength is higher than that of aluminum, and its corrosion resistance is comparable to that of platinum. This makes Ti very attractive for numerous applications in industry, technology, and science. The limiting factor, however, is its unfortunately difficult and expensive production.

#### 1.3.2.2 Chemical properties

Titanium reacts with air, chlorine, and other gases at relatively low temperatures. Compact Ti preserves its metallic luster up to 350 °C. With hydrogen it forms a hydride on heating. As confirmed by the negative standard potential of $Ti^{2+}$, Ti ($E_\Theta = -1.63$ V), titanium is a base metal. It is quite resistant to dilute and concentrated acids because of its passivation. Thus, it is not attacked by mineral acids, concentrated nitric acid, aqua regia (chloronitrous acid), and aqueous alkali at room temperature. In the presence of complexing agents, such as HF and oxalic acid, or at higher temperatures, dissolution takes place. With most nonmetals, titanium tends to form interstitial compounds, for example, TiN.

#### 1.3.2.3 Applications

Titanium metal is used in chemical apparatus construction (high resistance to corrosion), centrifuges, rotors, high vacuum technology, and electronic industry.

### Titanium(IV) compounds

#### Halides

Titanium halides are important starting compounds for various syntheses. Their properties vary considerably with the ON of titanium. The binary $TiX_4$ (X = Cl, Br, I) possesses a tetrahedral structure with polar covalent Ti–X bonds. $TiBr_4$ and $TiI_4$ are solids, and $TiCl_4$ is a colorless, fuming liquid. The industrial production of $TiCl_4$ from $TiO_2$, chlorine, and graphite has already been mentioned in the previous subsection. The following methods are more suitable for laboratory preparations:

$$TiO_2 + 2CHCl_3 \xrightarrow{500\,°C} TiCl_4 + 2CO + 2HCl$$

$$TiO_2 + CCl_4 \xrightarrow{500\,°C} TiCl_4 + 2CO_2$$

$$Ti + 2Cl_2 \longrightarrow TiCl_4$$

$TiCl_4$ and other halides hydrolyze rapidly forming an oxychloride and finally hydrated $TiO_2$:

$$TiCl_4 + H_2O \rightarrow TiOCl_2 + 2HCl$$

$$TiOCl_2 + 3H_2O \rightarrow TiO_2 \cdot 2H_2O + 2HCl$$

In the presence of hydrogen, reduction to purple $TiCl_3$ takes place:

$$TiCl_4 + H_2 \rightarrow 2TiCl_3 + 2HCl$$

$TiCl_4$ is soluble in various organic solvents, such as $CCl_4$, benzene, and hydrocarbons. Protic solvents, for example, acids and alcohol, cause protolysis. Numerous complexes with donor ligands have been described:

$$TiCl_4 + 2CH_3CN \rightarrow \left[TiCl_4(CH_3CN)_2\right]$$

#### Oxides

Titanium(IV) oxide, $TiO_2$, occurs in three modifications: anatase, brookite, and rutile. Brookite crystallizes in the orthorhombic system, whereas rutile and anatase possess a tetragonal crystal lattice. In all the three structures, titanium atoms are surrounded by six oxygen atoms and each O atom is coordinated by three Ti atoms. There are two industrially significant processes for the synthesis of $TiO_2$.

There are two industrially significant processes for the synthesis of $TiO_2$:

1. Sulfate process

$$FeTiO_3\ (\text{limonite}) \xrightarrow{H_2SO_4} FeSO_4 + Fe_2(SO_4)_3 + TiOSO_4$$

$$TiOSO_4 + H_2O\ (\text{excess}) \longrightarrow TiO_2 \cdot xH_2O + H_2SO_4$$

$$TiO_2 \cdot xH_2O \longrightarrow TiO_2 + xH_2O$$

2. Vapor phase oxidation of $TiCl_4$

$$TiCl_4 + O_2 \rightarrow TiO_2 + 2Cl_2$$

$TiO_2$ melts at 1,840 °C without decomposition and is inert to $H_2S$, acids, alkali solutions, and salt solutions. The reactivity of $TiO_2$ varies, however, considerably with the preparation conditions. At elevated temperatures $TiO_2$ reacts with $H_2$, CO, or carbon to give low-valent titanium oxides. Rutile is highly demanded for the production of titanium tetrachloride and organic titanium derivatives. Furthermore, it plays an outstanding role as white pigment dyestuff.

### Titanium(IV) acids and salts

The existence of a stoichiometric $Ti(OH)_4$ has not yet been proved. The precipitate obtained on adding a base to aqueous Ti(IV) solution is suggested to be the hydrated titanium(IV) oxide, $TiO_2{\cdot}2H_2O$, that is also called orthotitanic acid. On boiling, these hydrogels tend to coagulate under desorption of water eventually yielding metatitanic acid. The latter has the approximate formula $TiO_2{\cdot}H_2O$ or $Ti(OH)_2$, respectively. The exact composition of these $TiO_2{\cdot}xH_2O$ precipitates depends subtly on the pH value of the solution.

Titanic acids are amphoteric. They display only weakly basic character and are, therefore, soluble only in strong acids. In conc. acids, some Ti(IV) salts seem to be stable. In the presence of water, it hydrolyzes to hydroxo and oxo cations. Thus, according to the following protolysis equilibrium, under acidic conditions, the presence of octahedral hydroxo complex cations has been invoked:

$$Ti(OH)_4 + 3H_2O \rightarrow [Ti(OH)_3(OH_2)_3]^+ + OH^-$$

$$[Ti(OH)_3(OH_2)_3]^+ + H_2O \rightarrow [Ti(OH)_2(OH_2)_4]^{2+} + OH$$

The frequently employed titanyl cation, $TiO^{2+}$, is of some use to describe the reactions of Ti(IV) in aqueous solution, though it does not appear to exist either in solution or in the solid state. The crystalline salt $TiOSO_4{\cdot}H_2O$ contains the chain-like $(TiO)_n{}^{2n+}$ cation rather than a monomeric $TiO^{2+}$ species. Anhydrous $TiOSO_4$ can be easily synthesized by fusing $TiO_2$ with potassium pyrosulfate:

$$TiO_2 + K_2S_2O_7 \rightarrow TiO(SO_4) + K_2SO_4$$

On boiling or at higher pH values, titanyl salts tend to hydrolyze under separation of $TiO_2 \cdot H_2O$ or $TiO_2 \cdot 2H_2O$.

In accordance with its amphoteric behavior, $TiO_2$ dissolves in concentrated solution of alkali hydroxides to form hydrated titanates, such as $M_2{}^{I}TiO_3 \cdot nH_2O$ and $M_2{}^{I}Ti_2O_5 \cdot nH_2O$:

$$TiO_2 + 2OH^- \rightarrow TiO_3{}^{2-} + H_2O$$

Similar but anhydrous titanates are obtained by fusing the corresponding metal oxides with $TiO_2$.

Several minerals, for example, ilmenite, $FeTiO_3$, and perovskite, $CaTiO_3$, may also be classified as titanates. In addition to the so-called metatitanates, $M_2{}^{I}TiO_3$ and $M_2{}^{II}TiO_3$, orthotitanates, $M_4{}^{I}$ $TiO_4$, and $M_2{}^{II}TiO_4$, have also been described. Ti(IV) is capable of forming rather stable anionic complexes as well. Thus, on dissolving $TiCl_4$ in aqueous HCl oxo-chloro complex anions and with excess of HCl, the $[TiCl_6]^{2-}$ ion may be obtained. In analytical chemistry, the highly stable $[TiF6]^{2-}$ anion is of some interest.

### Peroxo complexes of titanium(IV)

The substitution of oxo groups by the peroxo anion in transition metal complexes is a facile reaction, usually yielding characteristically colored peroxo complexes. Thus, the $TiO^{2+}$ ion and related Ti(IV) species and $H_2O_2$ give the peroxotitanyl cation, $Ti(O_2)^{2+}$, in slightly acidic solution that has a typically orange color:

$$TiO_2 + \ + H_2O_2 \rightarrow Ti(O2)^{2+} + H_2O$$

## 1.4 Titanium(III) compounds

### Binary compounds

Titanium(III) halides are polymeric solids containing chloro-bridges. $TiCl_3$ occurs in four modifications. The purple α-form is obtained by reducing $TiCl_4$ with hydrogen at 500–1,200 °C. Low-temperature reduction of the tetrachloride with aluminum alkyls gives a fibrous β-$TiCl_3$, which is of industrial significance for the stereospecific polymerization of propylene. $TiCl_3$ forms numerous complexes, $[TiCl_3L_n]$ ($n = 1$–6), with suitable Lewis basic ligands L. $TiCl_3$ and other Ti(III) derivatives tend to disproportionate into Ti(IV) and Ti(II) on heating:

$$TiCl_3 \rightarrow TiCl_4 + TiCl_2$$

Ti(III) halides are strong reducing agents and are rapidly oxidized in air.

### Oxides and hydroxides

$Ti_2O_3$ is obtained from $TiO_2$ at 1,000 °C in a stream of hydrogen. It is quite inert and attacked only by oxidizing acids. Addition of aqueous alkali to Ti(III) salt solution precipitates dark purple $Ti(OH)_3$, which is most appropriately described as hydrous Ti(III) oxide. $Ti(OH)_3$ is rather sensitive to air and reduces even water under liberation of hydrogen.

### Aqueous chemistry of titanium(III)

Reduction of acidic Ti(IV) solutions by base metals, such as Mg or Zn, is the most common method to prepare Ti(III) compounds. These aqueous solution contains the violet $[Ti(H_2O)_6]^{3+}$ cation. Acidic Ti(III) solutions are also oxidized by air but at a much slower rate than the basic system. $Ti^{3+}$ solutions are of use in volumetric analysis as rapid and mild reducing agent.

The hexaquo ion $[Ti(H_2O)_6]^{3+}$ is the predominating constituent of slightly acidic solutions.

Nevertheless, it hydrolyzes partially but not to such an extent as Ti(IV) due to lower ON of the central atoms:

$$[Ti(H_2O)_6]^{3+} \rightarrow [Ti(OH)(H_2O)_5]^{2+} + H^+$$

From aqueous solutions of $Ti^{3+}$, two isomeric $TiCl_3{\cdot}6H_2O$ complex salts, one violet and the other brown, may be isolated. The type of isomerism encountered here is called hydrate isomerism.

## 1.5 Titanium(II) compounds

As Ti(II) has no aqueous chemistry, it is of minor interest for this practical course. For details, consult a comprehensive textbook of inorganic chemistry.

## 1.6 Chemicals and apparatus

| Chemicals | Apparatus |
|---|---|
| – Conc$^n$. sulfuric acid | Small flask |
| – Titanium powder | Water bath |
| – Trivalent titanium sulfate | Test tube |
| – Grain of zinc | |
| – Sodium hydroxide | |
| – Distilled water | |
| – Titanyl sulfate | |

(continued)

| Chemicals | Apparatus |
|---|---|
| – Dilute cupric chloride | |
| – Ferric chloride solution | |
| – Titanium(III) hydroxide | |

## 1.7 Procedure

### 1.7.1 Conversion of Ti metal to $Ti^{3+}$ by sulfuric acid

Fill the small flask three quarter full with semiconcentrated $H_2SO_4$, mix with a small amount of titanium powder, and heat the mixture for a few minutes in water bath. The start of the reaction is indicated by vigorous gas evolution. Later on, the solution turns violet due to the generation of the hexaquo cation $[Ti(H_2O)_6]^{3+}$. Save this solution for further tests:

$$2Ti + 6H^+ \rightarrow 2Ti^{3+} + 3H_2$$

### 1.7.2 Reduction of tetravalent titanium with zinc in dilute sulfuric acid

Place 12–15 drops of titanyl sulfate solution into a test tube and add a grain of zinc. Heat carefully until the solution becomes violet, the characteristic color of the $Ti^{3+}$ ion. Save the solution for further experiments:

$$Zn + 2H^+ \rightarrow Zn^{2+} + 2H$$

$$TiO^{2+} + H^+ \rightarrow H + Ti^{3+} + H_2O$$

### 1.7.3 Preparation of titanium(III) hydroxide

Add three to four drops of the $Ti^{3+}$ solution from Experiment 2.1 into another test tube and add four drops of 2 N NaOH. Do not shake the test tube in order to avoid side reactions. Bluish black titanium(III) hydroxide is precipitated:

$$Ti^{3+} + 3OH^- \rightarrow Ti(OH)_3(s)$$

### 1.7.4 Hydrolysis of titanium(III) and titanium(IV) salts

Place three to four drops of a solution of titanyl sulfate into one test tube and the same volume of the solution of trivalent titanium sulfate into another one (refer to Experiment 2.1). Add an equal volume of distilled water to both test tubes. Boil the solution for some minutes. In which case do you observe hydrolysis taking place? Explain your observation.

### 1.7.5 Reducing properties of titanium(III) hydroxide

Prepare $Ti(OH)_3$ as in Experiment 2.3. Do not shake the test tube. Observe the contents of the test tube carefully and you will see gas evolution. Moreover, the mixture slowly decolorizes and the blue $Ti(OH)_3$ precipitate turns white. This is because $Ti(OH)_3$ and other titanium(III) compounds are unstable to water in alkaline medium, and the following redox process takes place:

$$2Ti\,(OH)_3 + 2H_2O \rightarrow 2TiO_2 \cdot 2H_2O + H_2$$

### 1.7.6 Reducing properties of $Ti(OH)_3$: reaction with oxygen

Prepare $Ti(OH)_3$ as in Experiment 2.3. Then shake the test tube vigorously. The bluish precipitate turns white immediately without any gas evolution. This is due to the rapid oxidation of $Ti(OH)_3$ by atmospheric oxygen yielding $TiO_2 \cdot 2H_2O$:

$$4Ti(OH)_3 + O_2 + 2H_2O \rightarrow 4TiO_2 \cdot 2H_2O$$

In contrast to the above process, the reaction with water is much slower. Suggest an explanation.

### 1.7.7 Reducing properties of trivalent titanium: reduction of copper(II) chloride

Transfer three to four drops of the $Ti^{3+}$ solution prepared in Experiment 2.1 into another test tube and add two to three drops of dilute cupric chloride solution. The mixture is decolorized and a white precipitate of CuCl, which is insoluble in water, settles:

$$Ti^{3+} + Cu^{2+} + Cl^- + H_2O \rightarrow TiO^{2+} + 2H^+ + CuCl$$

### 1.7.8 Reduction properties of trivalent titanium: reduction of ferric chloride

Transfer three to four drops of the $Ti^{3+}$ solution prepared in Experiment 2.1 into another test tube and add three to four drops of ferric chloride solution. The violet color of $Ti^{3+}$ vanishes completely:

$$Ti^{3+} + Fe^{3+} + H_2O \rightarrow TiO^{2+} + Fe^{2+} + 2H^+$$

The resulting $Fe^{3+}$ is a pale green color that is barely recognizable when diluted.

## 1.8 Observation

Color change
Evolution of gas
Sound of reaction

## 1.9 Results and discussion

## 1.10 Review problems

1. Discuss two methods for the preparation of titanium metal from $TiO_2$.
2. Explain the ARKEL/De BOER process for the synthesis of pure titanium.
3. Describe the synthesis, structure, and properties of $TiCl_4$.

# Experiment 2

## 2.1 Chemistry of vanadium

## 2.2 Objective

Students should be able to:
- prepare vanadium pentoxide,
- observe the characteristic of vanadium pentoxide,
- observe the amphoteric property of $V_2O_5$,
- synthesize ammonium thiocyanate and vanadium pentasulfide, and
- identify vanadium(V) by the reaction of peroxovanadium(V).

## 2.3 Theory

### Occurrence

Vanadium is widespread in the Earth's crust (0.016%), but hardly found in great deposit. Compared to copper (0.01%) and cobalt (0.012%), vanadium cannot be considered a rare element. Some of the most frequent minerals are patronite, $V_2S_5{\cdot}XS$, vanadite, $3Pb_3(VO_4)_2{\cdot}PbCl_2$, and Carnotite, $K_2(VO_2)_2(VO_4)_2{\cdot}3H_2O$. In all these minerals, vanadium prefers the oxidation state +5. Compounds with vanadium in the +4, +3, and +2 oxidation states have been isolated from aqueous solution but do not occur in nature.

Vanadium has two isotopes: $^{50}V$ (0.24%) and $^{51}V$ (99.76%).

### Preparation

Most vanadium minerals do not contain more than 0.2% of the element. That is why, in the first step, they have to be concentrated up to 5% V. For further upgrading, the resulting slag is molten with $Na_2CO_3$ at 8,500 °C and, after cooling, extracted with water. On acidifying the aqueous vanadate extract, $V_2O_5$ is precipitated. Vanadium pentoxide itself or $VCl_3$ as another possible precursor can be reduced by base metals to give pure vanadium metal. Carbon and hydrogen are not suitable as reducing agents because they tend to form interstitial compounds with vanadium:

$$V_2O_5 + 5Ca \xrightarrow{9{,}000\,^{\circ}C} 5CaO + 2V \quad \text{(Van Arkel/De Boer process)}$$

$$VCl_3 + 3Mg \xrightarrow[\text{argon}]{800\,^{\circ}C} 3MgCl_2 + 2V \quad \text{(Van Arkel/De Boer process)}$$

Commercially, vanadium is mainly produced as an iron alloy called ferrovanadium.

https://doi.org/10.1515/9783111574349-003

## Properties

### Physical properties

Pure vanadium crystallizes in a cubic body-centered lattice with a melting point of 1,700 °C. Addition of carbon (interstitial) raises the melting temperature markedly up to 2,700 °C (10% carbon content).

The metal resembles titanium in being resistant to corrosion, hard, and steel-gray. Extremely pure vanadium is, ductile.

Table 1.4: Some physical properties of chromium.

| Properties | Z | Electronic configuration | Atomic weight | Density | m.p. | b.p. | χ | Atomic radius |
|---|---|---|---|---|---|---|---|---|
| **Expression** | 23 | $[Ar]3d^3 4s^2$ | 50.9414 | 7.19 g/c $m^3$ | 1,700 °C | 2,671 °C | 1.66 | 128 pm |

### Chemical properties

In the massive state, vanadium is not attacked by air, water, alkalis, or nonoxidizing acids, except HF. Oxidizing acids such as concentrated sulfuric acid and nitric acid dissolve the metal. At elevated temperature, it reacts with many nonmetals, for example, C, N, H, P, and As, mostly forming interstitial and nonstoichiometric compounds. Depending on the reaction temperature, various oxides have been observed, for instance, $V_2O_3$ (brown), $VO_2$ (black), and $V_2O_5$ (orange). Vanadium and chlorine react at 200 °C, yielding $VCl_4$ as a dark brown liquid.

### Oxidation state

The simple redox chemistry of vanadium is particularly interesting to inorganic chemists because vanadium readily exists in four different oxidation states: +5, +4, +3, and +2, corresponding to $d^0$, $d^1$, $d^2$, and $d^3$ electronic configurations.

### Application

About 95% of vanadium produced is consumed by steel industry for making hard and expansible sorts of steel. Vanadium pentoxide, ammonium metavanadate, and silver vanadate are of importance as catalysts in chemical industry ($SO_2$ oxidation, petrol cracking, etc.).

### 2.3.1 Vanadium(V) compounds

#### Halides

The only binary vanadium halide known is $VF_5$, a colorless, viscous liquid with polymeric structure. Furthermore, there are some oxyhalides $VOX_3$, (X = F, Cl, Br) and $VO_2X$ (X = F, Cl). All these halides are very sensitive to hydrolysis and not very stable to reduction.

#### Oxides

The most important vanadium oxide, $V_2O_5$, is usually prepared by thermal decomposition of ammonium metavanadate:

$$2NH_4VO_3 \xrightarrow{\text{heat}} V_2O_5 + 2NH_3 + H_2O$$

Another method of preparation is the precipitation from metavanadate solutions by dilute sulfuric acid:

$$2VO_3^- + 2H^+ \longrightarrow V_2O_5 + H_2O$$

$V_2O_5$ is a highly poisonous, brick red powder. It is sparingly soluble in water, yielding a slightly acidic, pale yellow solution. Although the acidic properties predominate, $V_2O_5$ also dissolves in acids. At higher temperature, $V_2O_5$ releases oxygen forming the dark blue $VO_2$. The latter can be reoxidized by amphoteric oxygen to $V_2O_5$. This particular property makes it an excellent catalyst for many oxidation reactions, for example, the contact process for sulfuric acid production and alcohol oxidation:

$$2VO_2 + O_2 \underset{\text{lower temp}}{\overset{\text{higher temp}}{\rightleftharpoons}} V_2O_5$$

#### Vanadate(V)

Vanadium pentoxide dissolves in aqueous alkali to give the colorless orthovanadate, $VO_4^{3-}$. This anion is stable only at pH 13:

$$V_2O_5 + 6OH^- \longrightarrow 2VO_4^{3-} + 3H_2O$$

Lowering of the pH value initiates polycondensation processes and the formation of so-called isopolyanions. Thus, in the pH range 13–3, the yellow $V_4O_7^{4-}$ is encountered. Between pH 8 and 6 the tetravanadate, $V_4O_{12}^{4-}$, and in acidic medium of pH 6–4 the decavanadate, $V_{10}O_{28}^{6-}$, were detected. The latter two anions have a bright orange color. In addition to various isopolyvanadates, there are also protonated species that may be considered as derivatives of the corresponding isopolyvanadic acids. At around pH 2, $V_2O_5$ is precipitated with more acids. It is redissolved to give the pale yellow vanadyl cation,$VO_2^+$:

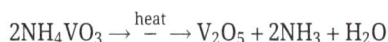

$$VO_4^{3-} + 2H^+ \rightleftharpoons VO_3^- + H_2O \quad \text{(isopoly vanadates)}$$

$$2VO_3^- + 2H^+ \rightleftharpoons V_2O_5 + H_2O$$

$$V_2O_5 + 2H \rightleftharpoons 2VO_2^+ + H_2O$$

The most common and commercially available vanadate is the tetravanadate, $Na_4$ $[H_2V_4O_{13}]$, that is conventionally named sodium metavanadate, $NaVO_3{\cdot}H_2O$. It is only the mentioned here that vanadium(V) is known to form various coordination compounds, which, in principle, can be regarded as derivatives of the ions $VO_2^+$ and $VO^{3+}$.

### Peroxo compounds

In addition to the vanadyl cation, $VO_2^+$, acid solutions may also contain the hydrated $VO^{3+}$ species. The latter has been reported to reacted with hydrogen peroxide to give the red-brown peroxovanadyl cation, $[V(O_2)]^{3+}$. Excess of $H_2O_2$ must be avoided because in the second step the pale yellow peroxo acid $H_3[VO_2(O_2)_2]$ is formed. This compound has the same color as $VO_2^+$:

$$VO^{3+} + H_2O_2 \longrightarrow V(O_2)^{3+} + H_2O$$

$$[V(O_2)]^{3+} + H_2O_2 + 2H_2O \longrightarrow H_3\left[VO_2(O_2)_2\right] + 3H^+$$

### Vanadium sulfides

Addition of ammonium sulfide to a neutral vanadium solution turns the color brown to red-violet due to the formation of thiovanadates, $VS_4^{3-}$:

$$VO_2^+ + 4S^{2-} + 4NH_4^+ \longrightarrow VS_4^{3-} + 4NH_3 + 2H_2O$$

On acidifying this solution, brown $V_2S_5$ is separated. Acidic V(V) solutions do not give a sulfide precipitation after $H_2S$ treatment but reduced to blue V(IV):

$$8VO_2^+ + H_2S + 14H^+ \longrightarrow 8VO^{2+} + S_4O^{2-} + 8H_2O$$

## 2.3.2 Vanadium(IV) compounds

### Halides

Where $VI_4$ has only been detected in the gas phase and $VBr_4$ decomposes already at −23 °C, the red-brown, monomeric liquid $VCl_4$ is quite stable and boils at 154 °C. Then latter is conveniently synthesized by chlorination of vanadium or $V_2O_5$:

$$V + 2Cl_2 \xrightarrow{270\ °C} VCl_4$$

$$V_2O_5 + 4CCl_4 \xrightarrow{\text{red hot}} VCl_4 + 4COCl_2 + \frac{1}{2}O_2$$

$VCl_4$ is very sensitive to hydrolysis and reacts violently with water to give a solution of various oxo-vanadium(IV) chlorides.

### Oxo-vanadium(IV) compounds

In case of vanadium oxidation state +4, which seems to be the most stable under normal condition by simple heating or with mild reducing agents, such as oxalic acid, $V_2O_5$ is converted into the dark blue $VO_2$. This oxide is amphoteric and soluble in acids and bases yielding either the blue $VO^{2+}$ cation or complex vanadate(IV), $VO_4^{4-}$.

The vanadyl cation, $VO^{2+}$, is actually hydrated and better described as $[VO(H_2O)_5]^{2+}$. Addition of alkali leads to the precipitation of yellow $VO(OH)_2$. The hydroxide is similarly amphoteric as the oxide and dissolves in acids and bases.

The $VO^{2+}$ cation can also be obtained by the reduction of $V_2O_5$ or $V_2O^+$ with $S_2O$ or HCl in acid aqueous solution.

The vanadyl(VI) cation has been shown to create an extensive coordination chemistry with complexes $[VOL_n]^{m\pm}$ ($n = 4$ or 5; L = neutral or anionic ligand).

## 2.3.3 Vanadium(III) compounds

### Halides

The trihalides $VX_3$ are shown for X being F, Cl, Br, and I. They are crystalline polymers with vanadium being octahedrally surrounded by six halogen atoms. The compounds are sensitive to hydrolysis and oxidized by air. On heating, they tend to disproportionate into $VX_2$ and $VX_4$:

$$2VX_3 \longrightarrow VX_2 + VX_4$$

The red-violet $VCl_3$ can be easily obtained by oxidative chlorination of vanadium, chlorination of vanadium, chlorination of $V_2O_3$, or by thermal decomposition of $VCl_4$:

$$VCl \xrightarrow{150\,°C} VCl_3 + \frac{1}{2}Cl_2$$

$$V_2O_3 + 3SOCl_2 \xrightarrow{200\,°C} 2VCl_3 + 3SO_2$$

$$2V + 6ICl \xrightarrow{\text{reflux}} 2VCl_3 + 3I_2$$

In water, all the halides hydrolyze to give the green hexaqua cation $[V(H_2O)_6]^{3+}$.

### Oxide and hydroxide

Vanadium oxide is a black solid prepared by reduction of $V_2O_5$ with hydrogen or carbon monoxide. It tends to form oxygen-deficient forms without structural changes. $V_2O_3$ is purely basic and dissolves in acids under formation of the $V^{3+}$ cation. Subsequent addition of alkali gives sparingly soluble, hydrated hydroxide, $V(OH)_3$. The latter is very sensitive to oxidation by air. Notice that oxidation is generally encouraged by a basic reaction medium.

### Vanadium(III) in aqueous solution

The green $[V(H_2O)_6]^{3+}$ and related ions can be prepared as mentioned above or by chemical and electrochemical reduction of vanadium compounds in the +4 and +5 oxidation states. The reduction of $VO_2^+$ by iodide in acidic medium is described in detail in the experiment part. $V^{3+}$ is also subject to aerial oxidation but considerably more stable than the hydroxide.

The hexaquo cation, $[V(H_2O)_6]^{3+}$, hydrolyzes partially in water to give hydrated $[V(OH)]^{2+}$ and $VO^+$ cations. It is, however, stable enough to occur in simple V(III) salts such as vanadium alums, $M^IV(SO_4)_2{\cdot}12H_2O$.

## 2.3.4 Vanadium(II) compounds

Vanadium(II) forms a black VO, which tends to nonstoichiometry, exhibits metallic properties, and is basic in nature. Therefore, it dissolves in acids.

Electrolytic or chemical reduction of vanadium compounds in higher oxidation state in acidic aqueous solution yields the violet, air-sensitive $[V(H_2O)_6]^{2+}$. $V^{2+}$ is a very powerful reducing agent, and reduces even water to hydrogen:

$$2V^{2+} + 2H_2 \longrightarrow 2V^{3+} + H_2 + 2OH^-$$

$[V(H_2O)_6]^{2+}$ is the constituent of several V(II) salts, for example, $VSO_4{\cdot}4H_2O$ and $M_2{}^I[V(H_2O)_6](SO_4)_2$ with M = $NH_4^+$, $K^+$, and so on. V(II) has a $d^3$ configuration. This is the reason for the kinetic inertness of $[V(H_2O)_6]^{2+}$ and related complexes in aqueous solution. Substitution reactions are evidently relatively slow.

## 2.4 Chemicals and apparatus

| Chemicals | Apparatus |
|---|---|
| – Aqueous solution of sodium metavanadate or with solid ammonium metavanadate ($NH_4VO_3$) | Spatula<br>Crucible |
| – Concentrated $H_2SO_4$ | Glass rod |
| – 2 N NaOH | Test tubes |
| – Distilled water | |
| – Sodium metavanadate | |
| – Barium chloride, copper sulfate, silver nitrate | |
| – Sodium or ammonium sulfide solution | |
| – 3% $H_2O_2$, oxalic acid, $H_2C_2O_4$ and $V_2O_5$ powder | |
| – Potassium iodide | |
| – Dilute solution of permanganate | |

## 2.5 Procedure

### 1 Preparation and reaction of vanadium compounds

#### 1.1 Preparation of vanadium pentoxide, $V_2O_5$

Heat a micro-spatula full of dry ammonium metavanadate, $NH_4VO_3$, in crucible with stirring.

Using glass rod, the white color of $NH_4VO_3$ will turn to brick red, which is the characteristic of $V_2O_5$:

$$NH_4VO_3 + \xrightarrow{\text{Heat}} V_2O_5 + 2NH_3 + H_2$$

#### 1.2 Amphoteric property of $V_2O_5$

Add very small amount of $V_2O_5$ (~0.1 mg) in each of three test tubes. Then add four to five drops of concentrated $H_2SO_4$ into the first test tube, equal amount of 2 N NaOH to the second test tube, and five to seven drops of distilled water into the third test tube and heat them in water bath for about 10 min. Observe color changes in each test tube:

$$V_2O_5 + 2H^+ \longrightarrow 2VO_2{}^+ + H_2O$$

$$V_2O_5 + 6\,OH^- \longrightarrow 2VO_4{}^{3-} + 3H_2O$$

Transfer a few drops from aqua solution of $V_2O_5$ into another test tube containing neutral litmus paper. Observe the change of color in litmus paper and select the possible equation from the following and explain:

$$V_2O_5 + H_2O \rightarrow 2VO_3^- + 2H^+$$

$$V_2O_5 + 3H_2O \rightarrow 2VO + 6H^+$$

$$V_2O_5 + H_2O \rightarrow 2VO_2^+ + 2OH$$

### 1.3 Conversion of tetravanadate to hexavanadate

Place three to four drops of sodium metavanadate (tetravanadate) into a test tube and add one drop of concentrated $H_2SO_4$. Pale yellow color of tetravanadate turns to yellowish orange, characteristics of hexavanadate, $V_6O_{17}^{4-}$, and related species. Write the ionic equation assuming metavanadate (a) as $VO_3^-$ and (b) $V_4O_3^-$

### 1.4 Preparation of sparingly soluble vanadate

Place three to four drops of sodium metavanadate (tetravanadate) into each of four test tubes and add the same amount of barium chloride, copper sulfate, silver nitrate, and lead acetate separately.

Depending on the pH of the medium, metavanadate or orthovanadate can be obtained. Write the color of each precipitate:

$$Ba^{2+} + 2VO_3^- \rightarrow Ba(VO_3)_2$$

$$Cu^{2+} + 2VO_3^- \rightarrow Cu(VO_3)_2$$

$$Pb^{2+} + 2VO_3^- + 2H_2O + 4Ac^- \rightarrow Pb(VO_4)_2 + 4HAc\ (Ac^- = CH_3COO^-)$$

$$Ag^{2+} + 2VO_3^- \rightarrow Ag(VO_3)$$

### 1.5 Synthesis of ammonium thiocyanate and vanadium pentasulfide

Place three to four drops of sodium metavanadate (tetravanadate) into a test tube and add dropwise sodium or ammonium sulfide solution until you observe brown or red-violet color depending on the sulfur content.

**Note:** Thiovanadate can be formed in neutral or slightly alkaline $(NH_4)_2S$ saturate with $H_2S$ to give very good $VS_4^{3-}$:

$$V_2O_5 + 4S^{2-} + 3H_2O \rightarrow VS_4^{3-} + 6OH^-$$

To the above solution, add 2N, HCl dropwise and you observe brown precipitate that indicates the formation of $V_2S_5$.

**Note:** Since $H_2S$ liberates on the reaction condition, some vanadium (V) reduces to soluble $VO^{2+}$ form. As a result, the solution turns to bluish:

$$VS_4^{3-} + 8H^+ \rightarrow V_2S_5 + 2H_2S$$

$$V_2S_5 + H_2S + 2H_2O \rightarrow VO^{2+} + S + SH^-$$

### 1.6 Identification of vanadium (V) by the reaction of peroxovanadium(V)

Place four to five drops of saturated sodium metavanadate into a test tube and add two to three drops of 2N, $H_2SO_4$ followed by one to two drops of 3% $H_2O_2$. Observe the color changes.

**Note:** A type of per-oxovanadium compound formed depends on the amount of $H_2O_2$, with limited amount of $H_2O_2$. $[VO(O_2)]^+$ is formed and such species have red-brown color; however, if excess $H_2O_2$ is added, the color will be pale yellow due to formation of $H_3[VO_2(O_2)_2]$. However, per-oxo compounds are unstable, and $H_2O_2$ decomposes due to the catalytic activity of transition element vanadium so that you will notice vigorous evolution of oxygen, $O_2$.

## 2 Preparation and reaction of vanadium(IV) compounds

### 2.1 Preparation of vanadium dioxide, $VO_2$

Mix equal amounts of oxalic acid, $H_2C_2O_4$, and $V_2O_5$ powder in a test tube. Fix the tube with an iron stand and heat. You will observe the dark blue color because of the formation of $VO_2$:

$$V_2O_5 + (COOH)_2 \xrightarrow{\text{heat}} VO_2 + CO_2 + H_2O$$

### 2.2 Amphoteric property of $VO_2$

From Experiment 2.1, place small amount of $VO_2$ into two test tubes and add four to five drops of 2N, NaOH into one of the tubes and equal amount of 2N, $H_2SO_4$ into the other. Heat both test tubes gently to dissolve the solid and observe the color of the resulting solution:

$$VO_2 + 2OH^- VO_3{}^{2-} + H_2O \left(\text{or } VO_4{}^{4-} \text{ and } V_4O_9{}^{2-}\right)$$

$$VO_2 + 2H^+ VO^{2+} + H_2O\,(\text{more precisely written as penta hydrated})$$

### 2.3 Reduction of vanadium(V) by sodium sulfite, $Na_2SO_3$, in aqueous solution

Place five to eight drops of saturated sodium metavanadate into a test tube and add two to three drops of 2N, $H_2SO_4$ followed by two to three crystals of $\mathbf{Na_2SO_3}$. Heat with stirring using glass rod until evolution of $SO_2$ gas stops. You will observe blue colored solution that contains $VO^{2+}$ and keep it for the next experiments:

$$VO_3{}^- + SO_2 + 4H^+ \rightarrow VO^{2+} + SO_4{}^{2-} + 2H_2O$$

Write another possible equation using $VO^+$ and $HSO_3{}^-$ as starting reactants.

### 2.4 Hydroxides of tetravalent vanadium

To the solution of Experiment 2.3, add 2N, NaOH dropwise and you will observe yellow precipitate due to the formation of $VO(OH)_2$:

$$VO^{2+} + 2OH^- \rightarrow VO(OH)_2$$

The hydroxide $VO(OH)_2$ is expected to be amphoteric so check it using 2N, NaOH and 2N, $H_2SO_4$ and write the equations.

### 2.5 Reduction property of tetravalent vanadium: reduction of $MnO_4^-$ in acidic media

Place three to five drops of dilute solution of permanganate, $MnO_4^-$, and add one to two drops of 2N, $H_2SO_4$ followed by two to three drops of vanadyl, $VO^{2+}$, solution. The solution decolorizes due to the reduction of permanganate:

$$VO^{2+} + MnO_4^- + H_2O \rightarrow 5VO^{2+} + Mn^{2+} + 2H^+$$

### 2.6 Identification of vanadium by reduction with hydrochloric acid, HCl, and reoxidation with iron(III) chloride

Place three to four drops of saturated sodium metavanadate into a test tube and add two to three drops of concentrated HCl. Boil it until the appearance of blue color, which indicates $VO^{2+}$. Then add two to three drops of ferric chloride, $FeCl_3$, and the blue color turns to slightly yellow color. This confirms reoxidation of $VO^{2+}$ to $VO^{2+}$:

$$VO^{2+} + 2Cl^- + 4H + VO^{2+} + Cl^{2+}2H_2O$$

$$VO^{2+} + Fe^{3+} + H_2O \rightarrow VO^{2+} + Fe^{2+} + 2H^+$$

## 3 Compounds of tetravalent vanadium

### 3.1 Reduction of pentavalent vanadium to tetravalent

Mix 5–10 drops of saturated sodium metavanadate and one to two drops of 2N, $H_2SO_4$ into a test tube.

Then add five to six drops of potassium iodide (KI) solution and heat the mixture until you observe green color of $[V(H_2O)_6]^{3+}$.

Note: Green color does not appear unless all iodine has been evaporated through boiling. In order to check the presence of iodine, transfer a drop of sample solution in a test tube containing 8–10 drops of starch solution. Iodine forms the dark blue color. To remove iodine, put the remaining solution in an evaporating dish and heat gently on a lower flame by adding some drops of water if the volume of the solution becomes too small.

When all iodine evaporates, the solution turns to green:

$$VO^{2+} + 3I^{-} + 4H^{+} \rightarrow V^{3+} + \frac{3}{2}I_2 + 2H_2O$$

### 3.2 Preparation and properties of vanadium(III) hydroxides

Place two to three drops of $V^{3+}$ from Experiment 3.1 into two test tubes and add two to three drops of 2N, NaOH.

Notice the formation of precipitate, $V(OH)_3$. This hydroxide is purely basic; prove that by testing its behavior toward 2N, $H_2SO_4$ and NaOH. Write the chemical equations of precipitate and dissolution.

## 4 Compounds of divalent vanadium

### 4.1 Successive reduction $V^{5+}$ to $V^{2+}$

Acidify 12–15 drops of saturated sodium metavanadate with 4–5 drops of concentrated HCl into a test tube and add 2–3 zinc grains.

Note: The test tube should not be too small as $H_2$ evolves. Gradual color changes from yellow ($VO^{2+}$) to green due to the mixture of $VO^{2+}/VO^{2+}$ to blue ($VO^{2+}$), eventually green ($V^{3+}$) to violet ($V^{2+}$):

$$VO^{2+} + Zn + 4H^{+} \rightarrow VO^{2+} + Zn^{2+} + 2H_2O$$

$$VO^{2+} + Zn + 4H^{+} \rightarrow V^{3+} + Zn^{2+} + 2H_2O$$

$$V^{3+} + Zn \rightarrow V^{2+} + Zn^{2+}$$

### 4.2 Preparation and properties of vanadium(II)hydroxides

Place two to three drops of $V^{3+}$ from procedure 4.1 into two test tubes and both test tubes add three to four drops of 2N, NaOH until the formation of precipitate, $V(OH)_2$. This hydroxide is purely basic; prove by testing its behavior toward 2N, $H_2SO_4$ and NaOH. Write the chemical equations of precipitate and dissolution.

**Note:** In the test tube, $Zn^{2+}$ also exists and can form $Zn(OH)_2$; however, it has amphoteric property so that it dissolves in excess NaOH through forming $[Zn(OH)_3]^{-}$. The remaining nonsoluble part of the precipitate is expected to be $V(OH)_2$.

## 2.6 Observation

Color change
Evolution of gas
Sound of reaction

## 2.7 Results and discussion

## 2.8 Review problems

1. In which types of chemical compound does vanadium occur in nature?
2. How do you prepare elemental vanadium from $V_2O_5$?
3. Vanadium tetrachloride is a brown oily liquid and may be synthesized from $V/Cl_2$ or $V_2O_5/SOCl_2$. Write the correct equations.
4. Discuss the acid–base properties of V oxides you do know.

# Experiment 3

## 3.1 Chemistry of chromium

## 3.2 Objective

Students should be able to:
- understand the most important oxidation states of chromium,
- describe the preparation of ferrochromium and pure chromium,
- prepare chromium trioxide, and
- understand hydrolysis of Cr(III) salts in the presence of carbonate and ammonium sulfide.

## 3.3 Theory

### The element

Chromium is named after Greek word *chrōma*, meaning color, because many of its compounds are intensely colored. Chromium is a steely-gray, lustrous, hard, and brittle metal found in Group 6 in the periodic table.

Table 1.5: Some physical properties of chromium.

| Properties | Z | Electronic configuration | Atomic weight | Density | m.p. | b.p. | χ | Atomic radius |
|---|---|---|---|---|---|---|---|---|
| **Expression** | 24 | $[Ar]3d^54s^1$ | 51.9961 | 7.19 g/c | 1,907 °C | 2,671 °C | 1.66 | 128 pm |

### Occurrence

In nature, chromium occurs as chromite (chrome iron ore), FeCr2O4, and crocoites (red lead ore), PbCrO4. Chromium is a trace element. Its content in the Earth's crust amounts to 200 g/t. The chromium content in the plant has been estimated to be 5 g/t. The most stable oxidation state of chromium is +3 although +7 is also quite common. Compounds with chromium in all the other positive ON have been synthesized as well.

### Preparation

If chromium is needed as an additive for iron alloys, chromite is directly reduced by carbon to give a carbon-containing ferroalloy called ferrochromate (about 60% chromium):

$$FeCr_2O_4 + 4C \rightarrow Fe/2Cr + 4CO$$

https://doi.org/10.1515/9783111574349-004

Pure chromium can be obtained by fusing chromite and alkali hydroxide in the presence of oxygen. Chromium is oxidized yielding soluble chromate(IV), whereas iron remains as insoluble Fe(OH)3. Chromate is then transferred into sodium dichromate followed by reduction with carbon to give chromium(III) oxide. Further reduction of Cr2O3 to metallic chromium is accomplished using aluminum as the reducing agent:

$$4FeCr_2O_4 + 7O_2 + 16NaOH \rightarrow 8Na_2CrO_4 + 4Fe(OH)_3 + 2H_2O$$

$$Na_2Cr_2O_7 + 2C \rightarrow Cr_2O_3 + Na_2CO_3 + CO$$

$$Cr_2O_3 + 2Al \rightarrow Al_2O_3 + 2Cr$$

## Properties

### Physical properties

Chromium is white shiny hard and brittle metal with the melting point of 1,903 ± 10 °C.

### Chemical properties

Chromium is insoluble in aqua regia (chloronitrous acid) and dilute or concentrated nitric acid because of its passivation . Nonoxidized mineral acids, such as HCl and $H_2SO_4$, dissolve the metal readily. At higher temperatures, chromium reacts with B, $N_2$, P, $O_2$, and some other nonmetals.

Especially Cr(VI) components but, to some extent, also other chromium-containing substances are highly carcinogenic.

### Application

Chromium is used in the industry for the production of steel. With Cr content higher than 12% steel becomes extraordinarily resistant to corrosion and heat because of passivating influence of particular elements. Chromium may serve as protective for other metals (chromium plating), and is also used for several alloys, as well as dying and tanning.

## 3.3.1 Chromium(II) compounds

### Binary chromium(II) compounds

Anhydrous Cr(II) halides, $Crx_2$ (x = Cl, B, I), are obtained by the action of HF, HCl, HBr, or iodine on the metal at 600–700 °C, or by reduction of the trihalides $CrX_3$ with hydrogen at 500–600 °C. Most important is the white, crystalline $CrCl_2$. All chromous halides dissolve readily in water to give the sky-blue hexaquo cation $[Cr(H_2O)_6]^{2+}$. In general, chromium(II) compounds are highly sensitive to oxygen.

#### The chromous cation and chromium(II) compounds

In aqueous medium, chromium(II) compounds may be synthesized either by reduction of Cr(III) precursor zinc amalgam or electrolytically as well as by dissociation of the metal in mineral acid under exclusion of air:

$$2Cr^{3+} + Zn \longrightarrow 2Cr^{2+} + Zn^{2+}$$

$$Cr + 2H^{+} \longrightarrow Cr^{2+} + H_2$$

Chromium(II) represents a $d^4$ system and forms many octahedral complexes. Most characteristic is a blue chromous cation $[Cr(H_2O)_6]^{2+}$. It possesses a high-spin electronic configuration $(t_{2g})^3(e_g)^1$. As the $e_g$ electron is easily slightly antibonding, it is readily split off under oxidation to Cr(III). Thus, $Cr^{2+}$ acts as a strong reducing agent:

$$4Cr^{2+} + O_2 + 4H^{+} \longrightarrow 4Cr^{3+} + 2H_2O$$

In the absence of suitable oxidizing agent, Cr(II) salts may even reduce water:

$$2CrCl_2 + 2H_2O \longrightarrow 2Cr(OH)Cl_2 + H_2$$

On addition of alkali hydroxide to an aqueous Cr(II) solution, the very air-sensitive, yellow $Cr(OH)_2$ is precipitated. This hydroxide is purely basic in nature.

### 3.3.2 Chromium(III) compounds

Of all common oxidation states of chromium, Cr(III) is assumed to be the most stable one.

#### Halides, $CrX_3$

$CrCl_3$ is a common starting material for synthetic purpose and can be prepared by:

a) Chlorination of the metal

$$Cr + \frac{3}{2}Cl_2 \longrightarrow CrCl_3$$

b) Dehydration of the hydrated chloride

$$CrCl_3 \cdot 3H_2O + 3SOCl_2 \longrightarrow CrCl_3 + 3SO_2 + 6HCl$$

c) De-oxygenation of the oxide

$$Cr_2O_3 + 3CCl_4 \longrightarrow 2CrCl_3 + 3COCl_2$$

Due to kinetic hindrance, $CrCl_3$ dissolves in water at an extremely low rate. Interestingly, the rate of dissolution can be enormously increased by catalytic quantities of $Cr^{2+}$. The latter is formed by adding traces of reducing agents such as Zn and Mg.

Chromium(III) is considered the most stable state of this element. In basic medium, especially to hexavalent, chromate occurs with oxidizing agents such as hydrogen peroxide and bromine. In acidic solution, such a powerful oxidant as peroxodisulfate is required for the preparation of dichromate, $Cr_2O_7^{2-}$, from chromium(III) compounds (of next subsection and experimental part).

### 3.3.3 Chromium(VI) compounds

$$CrO_2 + 2HCl \longrightarrow Cr_2O_2Cl_2$$

$$K_2Cr_2O_7 + 4KCl + 3H_2SO_4 \longrightarrow 2CrO_2Cl_2 + 3K_2SO_4 + 3H_2O$$

The only oxohalide that is much important to be mentioned here is chromyl chloride, $CrO_2Cl_2$, a deep-red liquid with b.p. 117 °C. The compound is obtained either from chromium(VI) oxide with concentrated HCl or by concentrating with dichromate in alkali chloride and concentrated sulfuric acid.

The latter reaction is of analytical use as it proves to be the way of separating even small amount of chromium from a sample by converting it into chromyl chloride and subsequent distillation of volatile product.

$CrO_2Cl_2$ is a strong oxidizing agent and hydrolyzes immediately in water.

#### Other oxo-chromium(VI) compounds

Chromium(VI) oxide, $CrO_3$, an orange, crystalline substance of m.p. 170 °C, precipitates on adding concentrated sulfuric acid to a saturated solution of alkali dichromate:

$$Cr_2O_7^{2-} + 2H^+ \longrightarrow 2CrO_3 + H_2$$

As $CrO_3$ and related Cr(VI) compounds are very efficient oxidizing agents, a solution of $K_2CrO_7$ in concentrated $H_2SO_4$, called chrome sulfuric acid, is frequently used for degreasing and cleaning laboratory glassware. The red-brown solution forms needles of $CrO_3$ on standing. A brown precipitate, casually settling in the same mixture, is chromyl sulfate, $CrO_2SO_4$:

$$CrO_3 + H_2O_4 \longrightarrow CrO_2SO_4 + H_2O$$

$CrO_3$ behaves like a typical acid anhydride, is soluble in water, and is very poisonous. Above its melting point, $CrO_3$ loses oxygen under formation of dark green $Cr_2O_3$. Chromite(VI), $M_2^ICrO_4$ (yellow), and dichromate(VI), $M_2^ICr_2O_7$ (orange), are usually synthesized by the oxidation of Cr(III) compounds in basic or acidic medium. Sparingly soluble chromates precipitate with $Ba^{2+}$, $Pb^{2+}$, and $Ag^+$.

On acidifying a chromate solution, its color turns from yellow to orange due to the formation of dichromate. This condensation reaction is initiated by protonation of the chromate followed by losing water:

$$CrO_4^{2-} + H^+ \longrightarrow CrO_3(OH)^- 2CrO_3(OH)^- \longrightarrow Cr_2O_7^{2-} + H_2O$$

These equilibria are extremely pH dependent. Further addition of acid leads to isopoly anions, such as $Cr_3O_{10}^{2-}$, $Cr_4O_{13}^{2-}$, and, finally, solid $CrO_3$. Whereas some oxygen acids tend to form simple hydrogen salts in acidified aqueous solution:

$$PO_4^{3} + H^+ \rightleftharpoons HPO_4^{2-}$$

$$Co_3^{2-} + H^+ \rightleftharpoons HCO_3^-$$

Elements like Cr, Mo, W, V, Nb, Ta, Si, and Sn are capable of condensing to polynuclear anions, that is, isopolyanion.

In contrast to basic chromate solutions, dichromate in acidic medium represents a strong oxidizing agent:

$$CrO_4^{2-} + 4H_2O + 3e^- \longrightarrow Cr(OH)_3 + 5OH^- \quad E^\theta = -0.13\ V$$

$$CrO^{2-} + 14HO^+ + 6e^- \longrightarrow 2Cr^{3+} + 21HO \quad E^\theta = +1.33\ V$$

Examples of oxidation with dichromate are given in the experimental part (e.g., $H_2S$, $I^-$, $Cl^-$, and $Fe^{2+}$). The following reaction is employed to detect alcohol in the breath of drunken car drivers:

$$\underset{(Orange)}{CrO^{2-} + 3CHOH + 8H^+} \longrightarrow \underset{(Green)}{2Cr^{3+} + 3CH\,CHO + 7HO}$$

## Peroxo compounds of chromium(VI)

In similarity with other transitional metals in higher oxidation state (e.g., Ti, V, Nb, Ta, Mo, and W), chromates form comparatively unstable but characteristically colored peroxo compounds. Thus, on treating acidic dichromate solutions with $H_2O_2$ at room temperature, deep blue $CrO_5$ is generated. The peroxide decomposes rapidly yielding green $Cr^{3+}$ and $O_2$. $CrO_5$ may, however, be extracted into ether and is then more stable:

$$CrO^{2-} + 4HO + 2H^+ \longrightarrow 2CrO(O) + 5HO$$

$$4CrO(O_2)_2 + 12H^+ \longrightarrow 4Cr^{3+} + 6H_2O + 7O_2$$

At temperature below 0 °C, green cationic peroxo species, such as $[Cr_2(O_2)]^{4+}$ and $[Cr_3(O_2)_2]^{5+}$, have been detected. From neutral or slightly acidic solutions of dichromate and $H_2O_2$, blue violet peroxo salts believed to contain the ion $[Cr(O_2)_2(OH)]^-$ could be separated.

Alkaline chromate solutions and 30% $H_2O_2$ give, under certain conditions, red-brown peroxochromates, $M_3^{I}CrO_8$. All solid peroxo salts are highly explosive.

## 3.4 Chemicals and apparatus

| Chemicals | Apparatus |
|---|---|
| Ammonium dichromate $(NH_4)_2Cr_2O_7$ | Test tube |
| Aqueous solution of $KrCr(SO_4)_2{\cdot}12H_2O$ | Iron stand |
| Chromium sulfate | Bunsen burner |
| 2N, NaOH and 2N, $H_2SO_4$ | Water bath |
| $Na_2CO_3$, $(NH_4)_2SO_4$,$CrO_3$ | |
| Potassium dichromate | |
| $CrCl_3{\cdot}6H_2O$ solid | |
| 3% $H_2O_2$ | |
| 2N nitric acid | |
| Potassium chromate ($K_2CrO_4$), $CrO_3$ | |

## 3.5 Procedure

### 1 Preparation and reaction of Cr(III) compounds

#### 1.1 Preparation of Cr(III) oxide, Cr2O3

Put $(NH_4)_2Cr_2O_7$ powder in a test tube and clam it vertically with iron stand and heat until the color of ammonium dichromate disappears. The appearance of dark green tells you the formation of chromium(III) oxide:

$$(NH_4)_2Cr_2O_7 \rightarrow Cr_2O_3 + N_2 + 4H_2O$$

#### 1.2 Preparation of Cr(III) hydroxide, Cr(OH)3

In two test tubes, place three to four drops of aqueous $KrCr(SO_4)_2{\cdot}12H_2O$ or chromium sulfate, $Cr_2(SO_2)_3$, and add one to two drops of 2N, NaOH into both tubes until you observe grayish green, gel-like precipitate, which confirms the formation of $Cr(OH)_3$.

**Note:**

1. Both $KrCr(SO_4)_2{\cdot}12H_2O$ and $Cr_2(SO_4)_3$ contain $[Cr(H_2O)_6]^{3+}$ complex ion
2. Addition of excess hydroxides leads to the disappearance of precipitate, which may be due to the formation of $[Cr(OH)_6]^{3-}$.

#### 1.3 Amphoteric property of Cr(OH)3

In one of the test tubes, refer Experiment 1.2, add dropwise 2N, $H_2SO_4$ and 2N, NaOH into the other and observe the color of the resulting solution:

$$Cr(OH)_3 + 2\ N\ 3H_2SO_4 \xrightarrow{\text{dropwise}} Cr^{3+} + 3H_2O$$

$$Cr(OH)_3 + 2\ N\ NaOH \xrightarrow{\text{dropwise}} [Cr(OH)_6]^{3-}$$

### 1.4 Hydrolysis of sodium hexahydroxochromate(III): refer 1.3. above

Heat hexahydroxochromate(III), $[Cr(OH)_6]^{3-}$, in a test tube in water bath until grayish color appears, which tells you the formation of Cr(III) hydroxide, $Cr(OH)_3$:

$$[Cr(OH)_6]^{3-} \xrightarrow{\Delta} Cr(OH)_3 + 3OH^-$$

The equilibrium will be shifted to the right due to heat.

### 1.5 Hydrolysis of Cr(III) salts

In five drops of litmus solution in a test tube, add two drops of chromium(III) sulfate or chromicōsulfate, $Cr_2(SO_4)_3 \cdot 12(H_2O)$ and observe the color change and write the chemical equation. If **violet-blue color** appears, then that indicates the presence of $Cr^{3+}$:

$$\text{Litmus (5d)} + Cr_2(SO_4)_3(2d) \rightarrow [Cr(H_2O)_6OH]^{3+} + H_3O^+$$

### 1.6 Hydrolysis of Cr(III) salts in the presence of carbonate and ammonium sulfide

Place three to four drops of aqueous solution chromic salt in each of two test tubes and add dropwise $Na_2CO_3$ solution in one of the tubes and $(NH_4)_2S$ into the other. You will observe greenish precipitate due to the formation of Cr(OH)3:

$$Cr^{3+}(3d) + Na_2CO_3(dw) \rightarrow Cr(OH)_3 + Na_2SO_4 + CO_2$$

$$Cr^{3+}(3d) + NH_4)_2S\ (dw) \rightarrow Cr(OH)_3 + Na_2SO_4 + H_2S \qquad (dw = dropwise)$$

**Note:** Carbonates and sulfides in water undergo protolysis, releasing gaseous $CO_2$ and $H_2S$ as byproducts.

$$CO_3{}^{2-} + H_2O \rightarrow HCO^{3-} + OH^-$$

$$HCO^{3-} + OH^- \rightarrow CO_2 + OH^-$$

$$S^{2-} + H_2O \rightarrow HS^- + OH^-$$

$$HS^- + H_2O \rightarrow H_2S + OH^-$$

$$Cr^{3+} + 3OH^- \rightarrow Cr(OH)_3$$

### 1.7 Aqua complexes of Cr(III)

In two separate test tubes, pour five to six drops of water and add some $CrCl_3{\cdot}6H_2O$ solid. Heat one of the test tubes in boiling water bath for not less than 10 min until the color of the solution changes. Compare the color with the test tube that has been kept at room temperature:

$$CrCl_3 \cdot 6H_2O\ (0.01\ g) + H_2O\ (6d) + heat \longrightarrow [Cr(H_2O)_4Cl_2]^{+} green$$

$$CrCl_3 \cdot 6H_2O\ (0.01\ g) + H_2O\ (6d) \longrightarrow CrCl_3 \cdot 6H_2O$$

**Note:** Chromic chloride has violet-blue color due to $[Cr(H_2O)_6]^{+}$; however, green color may arise from the formation of $[Cr(H_2O)_5Cl]^{+}$ (pale-green) or $[Cr(H_2O)_4Cl_2]^{+}$ (dark green).

## 2 Preparation and reaction of Cr(VI) compounds

### 2.1 Preparation of chromium trioxide, $CrO_3$

Dissolve 5 g of potassium dichromate in 10 mL of boiling water. Cool the solution to room temperature and very slowly add 7 mL of concentrated sulfuric acid. Allow it to settle down for 2 h, and then filter off the liquid potassium hydrogen sulfate from the crystals. Heat the filtrate to 85 °C and add 5 mL of dilute sulfuric acid.

Evaporate the liquid on a water bath until crystals form on this surface, then set it aside to crystallize. Filter through glass wool, preferably with suction, and evaporate the filtrate to produce more crystals. To remove traces of sulfuric acid, wash the crystals while still in the filter with concentrated nitric acid. Chromium trioxide is not soluble in nitric acid. Transfer the crystals to a dry evaporating basin and heat in an air oven at 130 °C:

$$K_2Cr_2O_7 + 2H_2SO_4 \longrightarrow 2KHSO_4 + 2CrO_3(s) + H_2O$$

### 2.2 Decomposition of chromic anhydride (chromium trioxide, $CrO_3$)

Place 0.05 g of chromium trioxide in a crucible and heat it with a Bunsen burner on asbestos wire gauze. The color changes from brown to dark green, which is the characteristic color of $Cr_2O_3$:

$$CrO_3(0.05\,g) + heat \longrightarrow Cr_2O_3 + \frac{3}{2}O_2$$

### 2.3 Oxidation of trivalent Cr to chromate, Cr(VI), by hydrogen peroxide in alkaline media

**Note:** Cr(III) can be oxidized to Cr(VI) in an acidic media but it needs extremely powerful oxidizing agents. **Refer Experiment 1.3.**

Place $[Cr(OH)_6]^{3-}$ in a test tube and add one to two drops of NaOH followed by addition of three to five drops of 3% $H_2O_2$. Then heat the mixture in a water bath until green color changes into yellow, which indicates the formation of chromate:

$$[Cr(OH)_6]^{3-} + NaOH + 3\%\ H_2O_2 \rightarrow CrO_4^{2-} + H_2O + 2OH^-$$

### 2.4 Oxidation of trivalent to Cr to chromate, Cr(VI), by peroxodisulfate, $K_2S_2O_8$

Place three to four drops of aqueous solution of chromium(III) salt in a test tube and add the same volume of 2N, $H_2SO_4$ followed by a few crystals of potassium peroxodisulfate and boil the mixture for a minute. The color of the mixture would change, which indicates the formation of $Cr2O_7^{2-}$:

$$Cr_3 + (\mathbf{3d}) + H_2SO_4(\mathbf{3d}) + S_2O_8^{2-}(1\,g) \rightarrow Cr_2O_7^{2-} + SO_4^{2-} + 14H^+$$

### 2.5 Interconversion of chromate and dichromate

$$2K_2CrO_4 + H_2SO_4 \rightarrow K_2SO_4 + K_2Cr_2O_7 + H_2O$$

$$2\,CrO_4^{2-}(\mathbf{3d}) + 2H^+(\mathbf{dw}) \rightarrow 2CrO_4^{2-} + H_2O$$

To three to four drops of potassium dichromate solution, add sodium hydroxide solution dropwise until its orange color changes to yellow, which conforms the formation of the chromate ion:

$$Cr_2O_7^{2-}(\mathbf{3d}) + 2OH^-(dw) \rightarrow 2CrO_4^{2-} + H_2O$$

### 2.6 Preparation of sparingly soluble chromates

Place three drops of potassium chromate solution into three separate test tubes. Then add equal volume of BaCl2 into the first test tube, lead acetate into the second test tube, and silver nitrate into the third test tube. Note the color changes and write the ionic equation:

$$2CrO_4^{2-}(\mathbf{3d}) + M^{n+}(\mathbf{3d}) \rightarrow M_2(CrO_4) \qquad (M = Ba^{2+}, Pb^{2+}, Ag^+)$$

### 2.7 Oxidation property of hexavalent chromium

#### 2.7.1 Oxidation of hydrogen sulfide, $H_2S$

Place five drops of potassium dichromate solution in a test tube and acidify it with three drops of 2N, $H_2SO_4$. Then add freshly prepared $H_2S$ dropwise until the color changes to green, which indicates the formation $Cr^{3+}$:

$$Cr_2O_7^{2-}(\mathbf{5d}) + H_2S\,(dw) + 8H^+(\mathbf{3d}) \rightarrow 2Cr^{3+} + 3S^+7H_2O$$

### 2.7.2 Oxidation of Iodide by dichromate in acidic media

Place five drops of potassium dichromate solution in a test tube and acidify it with three drops of 2N, $H_2SO_4$. Then add five drops of potassium iodide solution. Observe the color changes but if green color does not appear, heat the tube to avoid iodide ion:

$$Cr_2O_7^{2-}(\mathbf{5d}) + KI\,(4d) + 14H^+(\mathbf{4d}) \rightarrow 2Cr^{3+} + 3I_2 + 7H_2O$$

### 2.7.3 Oxidation of hydrochloric acid by dichromate

Place three drops of potassium dichromate solution, and add five drops of concentrated hydrochloric acid until orange color changes to green, which is characteristic for $Cr^{3+}$:

$$Cr_2O_7^{2-}(\mathbf{3d}) + 6\,Cl^-(\mathbf{4d}) + 14H^+(\mathbf{5d})\,2Cr^{3+} + 3Cl_2 + 7H_2O$$

### 2.8 Peroxo compounds of chromium(VI) – blue chromium peroxide

Place five drops of potassium dichromate solution, and add drops of 2N, nitric acid until orange color changes to green, which is the characteristic for $Cr^{3+}$. Then add 0.5 mL diethyl ether and three drops of 3% $H_2O_2$. Formation of blue color is the pointer for the formation of chromium peroxide:

$$Cr_2O_7^{2-}\ (\mathbf{3d}) + 2\ N\,6HNO_3(1d) + H_2O_2(\mathbf{3d}) \rightarrow 2HCrO_5 + 2KNO_3 + 7H_2O$$

$$Cr_2O_7^{2-}(\mathbf{3d}) + 2H^+ + 5H_2O_2(\mathbf{3d}) \rightarrow 2CrO_5 + 7H_2O$$

$$2CrO_5 \rightarrow Cr_2O_3 + O_2$$

## 3.6 Observation

Color change
Evolution of gas
Sound of reaction

## 3.7 Results and discussion

## 3.8 Review questions

1. Describe the preparation of ferrochromium and pure chromium.
2. $Cr_2O_3$ exhibits amphoteric properties. Find chemical reactions to confirm this.
3. Write the equation for the following reactions:
   a. Addition of NaOH to Cr(III) salt solutions
   b. Chromium(III) chloride with sodium sulfide in water

4. Write the Lewis formulae of:
   a. $CrO_2^-$; b. $CrO_4^-$; c. $Cr_2O_7^{2-}$
5. $CrCl_3$ exists in the form of three differently colored hydrate isomers. Write down their exact formulae.
6. Complete the following reaction equations:
   a. $Cr^{3+} + S_2O_8^{2-} + H^+ \longrightarrow$
   b. $Cr_2O_7^{2-} + I^- + H^+ \longrightarrow$
   c. $Cr_2O_7^{2-} + SO_3^{2-} + H^+ \longrightarrow$
   d. $Cr(OH)_3 + H^+ \longrightarrow$
   e. $Cr(OH)_3 + OH^- \longrightarrow$
   f. $Cr^{3+} + H_2 \longrightarrow$
   g. $Cr^{2+} + H_2O \longrightarrow$
   h. $CrO_4^{2-} + H^+ \longrightarrow$

# Experiment 4

## 4.1 Chemistry of molybdenum

## 4.2 Objective

Students should be able to:

- describe the preparation and properties of $MoCl_5$,
- identify Mo(V) in aqueous solution by means of two precipitation and two redox reactions, and
- understand which one is the most common oxidation states of molybdenum.

## 4.3 Theory

### 4.3.1 The element

Molybdenum is named molybdenum in Neo-Latin and molybdos in ancient Greek, meaning lead, since its ores were confused with lead ores. Silvery metal with a gray cast has the sixth highest melting point.

#### Occurrence

Molybdenum does not occur naturally as a free metal on the Earth, instead being found only in various oxidation states in minerals. Molybdenum is most frequently found in molybdenite, $MoS_2$, and to some extent in wulfenite, $PbMoO_4$. It is a trace element like that of chromium.

#### Preparation

Initially, sulfidic molybdenum ores are concentrated by froth floatation followed by roasting to give $MoO_3$. The oxide has to be further upgraded by dissolution in ammonia and calcination of the resulting ammonium molybdate. The $MoO_3$ thus purified can be reduced by hydrogen yielding elemental molybdenum:

$$MoS_2 + 3.5O_2 \rightarrow MoO_3$$

$$MoO_3 + 3H_2 \rightarrow Mo + 3H_2O$$

Ferromolybdenum, an alloy of iron and molybdenum, is manufactured electrothermally from a mixture of iron oxide, $MoO_3$, and coal.

https://doi.org/10.1515/9783111574349-005

## Properties

### Physical properties

In the powder form, Mo is dull gray. After fusion in the massive state, the metal has a lustrous, silver-white appearance. Its electrical conductance is about 30% that of silver. The metal melts at 2,610 °C.

Table 1.6: Some physical properties of molybdenum.

| Properties | Z | Electronic configuration | Atomic weight | Density | m.p. | b.p. | χ | Atomic radius |
|---|---|---|---|---|---|---|---|---|
| **Expression** | 42 | $[Ar]4d^5 5s^1$ | 95.95 | 10.28 g/c $m^3$ | 2,623 °C | 4,639 °C | 2.16 | 139 pm |

### Chemical properties

Molybdenum is inert to oxygen at room temperature but reacts to give $MoO_3$ at red heat. The metal is dissolved by concentrated nitric acid, boiling concentrated sulfuric acid, and chloronitric acid. Dilute and nonoxidizing acids do not attack the element.

### Application

Molybdenum is largely used for the production of various steel alloys to increase their hardness and toughness. Molybdenum filaments are needed for electrical bulbs. A great many industrial catalysts are based on $MoO_3$ and $MoS_2$ (e.g., petrol cracking, oxidation of benzene to phthalic anhydride, and metathesis of propene into ethane and butane). The sheet-like $MoS_2$ is very common grease-free lubricating agent. Molybdenum-containing enzymes play a significant role in the biological nitrogen fixation.

## 4.3.2 Molybdenum(VI) compounds

### Halide and oxyhalides

All molybdenum(VI) halide and oxyhalides are highly sensitive to moisture and they hydrolyze rapidly. The following list gives a brief impression of the chloro and fluoro derivatives:

$$MoFMo + 3F_2\ MoF_6 \longrightarrow \text{colorless, volatile, m.p. 17.5 °C, b.p. 35 °C}$$

$$MoOF_4Mo + 2F_2 + O \longrightarrow MoOF_2 \text{ colorless, volatile, m.p. 975 °C}$$

$$MoOCl_4 + 4HF\ MoOF_4 + 4HCl \longrightarrow \text{b.p. 180 °C}$$

$$MoO_2F_2\ MoF_6 + 2H_2O \longrightarrow MoO_2F_2 + 4HF \qquad \text{White solid sublimes at 270 °C}$$

$$MoO_2Cl_2 + 2HF \longrightarrow MoO_2F_2 + 2HCl \qquad (1.013\,bar)$$

$MoCl_6$ not existent

$$MoOCl_4MoCl_5 + \frac{1}{2}O_2 \longrightarrow MoOCl_4 + \frac{1}{2}Cl_2 \quad \text{Dark green crystals}$$

$$MoO_3 + 2SOCl_2 \longrightarrow MoOCl_4 + 2SO_2 \quad \text{m.p. 102 °C}$$

$$MoO_2Cl_2MoO_2 + Cl_2 \longrightarrow MoO_2Cl_2 \qquad \text{White crystals, m.p. 175 °C}$$

On acidifying aqueous molybdate solution by an excess of HCl, various chloro complexes are formed: 2N HCl: $[MoO_2Cl_2(H_2O)_2]$ and $[MoO_2Cl_3(H_2O)]^-$ 12N HCl:*cis* – $[MoO_2Cl_4]^{2-}$

## Oxide

Molybdenum trioxide, $MoO_3$, is a white solid turning yellow on heating (m.p. 795 °C). It crystallizes in a rare layer lattice with each molybdenum coordinated by six oxygen atoms in distorted octahedral manner.

The hydrate, $MoO_3{\cdot}XH_2O$ ($X = 1,2$), may be obtained as white precipitate on acidifying aqueous molybdate solutions:

$$MoO_4^{2-} + H_2O + 2H^+ \longrightarrow MoO_3 \cdot 2H_2O \; (\text{or } MoO_3 \cdot H_2O)$$

NMR spectroscopy has shown that these compounds are molybdenum(VI) oxide hydrate rather than molybdic acid.

## The molybdenyl cation and simple molybdates

Molybdenum oxide has amphoteric properties. Freshly prepared $MoO_3$ hydrates dissolve in strong acid to give the molybdenyl cation, $MoO_2^{2+}$:

$$MoO_3 + 2H^+ \longrightarrow MoO_2^{2+} + H_2O$$

An analytically interesting reaction is the precipitation of red-brown $(MoO_2)_2[Fe(CN)_6]$ from HCl-acidic Mo(VI) solution and ferrocyanide.

In alkali solutions, $MoO_3$ hydrates are dissolved under the formation of monomeric, colorless orthomolybdates, $MoO_4^{2-}$:

$$MoO_3 + 2OH^- \longrightarrow MoO^{2-} + HO$$

Only alkali, ammonium, and thallium(I) molybdates are soluble in water.

## Isopoly and heteropoly anions and acids

On acidifying a slightly basic molybdate solution, polycondensation under abstraction of water takes place, yielding a great variety of oligomeric and polymeric, colorless molybdate called isopolymolybdates. Finally, polymeric, hydrated molybdenum oxide, $MoO_3{\cdot}xH_2O$, is precipitated. The stability of different anions and the extent of polycondensation depend heavily on the pH value of the system:

$$2MoO + 2H^+ \rightleftharpoons Mo_2O_7^{2-} + H_2O$$

$$3Mo_2O_7 + 2H^+ \rightleftharpoons 2Mo_3O_{10}^{2-} + H_2O$$

$$6MoO_4^{2-} + 8H^+ \rightleftharpoons Mo_6O_{21}^{6-} + 2H_2O$$

$$Mo_6O_{21}^{6-} + 6H^+ \rightleftharpoons MoO_3 + 3H_2O$$

Isopoly acids and salts are readily formed with B, Si, Sn, P,V, Nb, Ta, Cr, Mo, and W. They are generated either by acidifying an aqueous solution of the monomeric precursor (the ortho anion) or by heating of a suitable solid salts. The polyanions may be linear or cyclic structures. If such polyanions involves two or more different central atoms, it is called a heteropoly anion or acid, respectively. These species are normally prepared by combining solutions of both a weak metal acid ($H_2WO_4$, $H_6Mo_6O_{21}$, etc.) or its anion and weak or moderate strong nonmetal acid or its anion. The latter can be $H_3BO_3$, $H_3PO_4$, $H_4SiO_4$, and so on. Many salts of these heteropoly acids are sparingly soluble and, therefore, of analytical interest. One of the most representative examples is the slightly soluble, yellow ammonium dodecylmolybdatophosphate hexahydrate:

$$2MoO^{6-} + 11H^+ + HPO^{2-} + 3NH^+ \longrightarrow (NH)[P(MoO)] \cdot 6H_2O$$

This precipitation reaction is usually employed for the qualitative and quantitative determination of molybdenum and phosphorus.

#### Sulfide and thiomolybdate

Black-brown $MoS_3$ is formed from molybdate and $H_2S$ in aqueous solution:

$$MoO_4^{2-} + 3H_2S \longrightarrow MoS_3 + 2H_2O + 2OH^-$$

The reaction proceeds very slowly and, incidentally, has to be carried out under $H_2S$ pressure. $MoS_3$ is sparingly soluble in concentrated HCl, but readily soluble in chloronitric acid, and yellow ammonium sulfide (ammonium polysulfide, $(NH_4)_2S_X$) solution. Molybdate and $(NH_4)_2S_X$ give red thiomolybdate, $MoS_4{}^{2-}$, which on acidifying is converted into insoluble $MoS_3$

### 4.3.3 Molybdenum(V) compounds

On heating stoichiometric amounts of $MoO_3$ and Mo or $MoO_2$, respectively, in evacuated quartz tube, some oxides with molybdenum in formal oxidation state between +5 and

+6 are formed. These oxides are nonstoichiometric substances with formulae such as $Mo_9O_{26}$ and $Mo_{18}O_{52}$.

They are closely related to the so-called molybdenum blue compounds. The latter are frequently described with formulae such as $Mo_4O_{10}(OH)_2$ or $Mo_3O_8$(=$MoO_3$, $Mo_2O_5$), and $Mo_2O_5$. Molybdenum blue can be obtained by partial reduction of $MoO_3$ or molybdate(VI) with Zn/HCl, $SnCl_2$/HCl, concentrated $H_2SO_4$, and other reducing agents (see experimental part). The reaction is a convenient confirmatory test for traces of molybdenum. Tungsten gives similarly blue oxides.

Molybdenum(V) hydroxide, $MoO(OH)_3$, precipitates on adding ammonia to an aqueous solution of a Mo(V) compound. A higher temperature of it dehydrates to give blue $Mo_2O_5$.

Molybdenum(V) chloride, $MoCl_5$, is another Mo(V) compound of great interest. The black, crystalline solid (m.p. 194 °C, b.p. 628 °C) proves to be extremely sensitive to hydrolysis. $MoCl_5$ is commercially available and readily obtained by heating molybdenum powder in a stream of chloride at 300–400 °C:

$$Mo + 25Cl_2 \longrightarrow MoCl_5$$

The chloride has high affinity to oxygen. Its dissolution in water or other oxygen-containing solvents usually results in the generation of green $MoOCl_5$ derivatives. A large variety of complexes with the moiety $MoOCl_3$ and related compounds have been described.

### 4.3.4 Molybdenum(VI) compounds

$MoO_2$ is generated by synproportionation from $MoO_3$ and Mo or by reduction of $MoO_3$ with hydrogen:

$$2MoO_3 + Mo \xrightarrow{1{,}000\ ^\circ C} 3MoO_2$$

$$MoO_3 + H_2 \longrightarrow MoO_2 + H_2O$$

$MoCl_4$ can be obtained as a blackish brown powder by refluxing $MoCl_5$ in benzene or other organic solvents:

$$2MoCl_5 + C_6H_6 \longrightarrow MoCl_4 + C_6H_5Cl + HCl$$

With various $O^-$, $N^-$, and p-donor ligands, complexes of composition $MoCl_4L_2$ may be synthesized.

### 4.3.5 Molybdenum(III) compounds

$Mo_2O_3$ is formed on reducing $MoO_3$ with potassium in liquid ammonia:

$$MoO_3 + 6K \longrightarrow 3Mo_2O_3 + 3K_2O$$

$MoCl_3$ was prepared from $MoCl_5$ and $SnCl_2$ at 300 °C:

$$2MoCl_5 + SnCl_2 \xrightarrow{300\,°C} MoCl_3 + SnC$$

Aqueous Mo(III) solution is synthesized by chemical or electrochemical reduction of molybdenum(VI) compounds. The common reducing agents for this purpose are $SnCl_2$, Zn, and $Na_2S_2O_3$ in acidic medium. In the presence of thiocyanate, $SCN^-$, the characteristic, red $[Mo(SCN)_6]^{3-}$ complex anion is formed. The latter can be used to identify molybdenum even in the presence of tungsten. Tungsten(VI) is only reduced to tungsten blue by these reducing agents.

## 4.4 Chemicals and apparatus

| Chemicals | Apparatus |
|---|---|
| Ammonium molybdate(VI), $(NH_4)_2MoO_4$ (aq) | Dropper |
| HCl, $H_2SO_4$, or $HNO_3$ | Test tube |
| 2N, NaOH and 2N, HCl | Water bath |
| Potassium hexacyanoferrate(II) (aq) | Bunsen burner |
| Calcium chloride and lead acetate | |
| Hydrogen phosphate (aq) | |
| Ammonium sulfide | |
| Ammonium molybdate (solid) | |
| 3% $H_2O_2$ and concentrated ammonia | |
| Stannous chloride and grains of zinc | |
| Potassium thiocyanate | |

## 4.5 Procedure

### 1 Preparation and reaction of Mo(VI) compounds

#### 1.1 Preparation and properties of molybdic acid, $MoO_3{\cdot}H_2O$ or $H_2MoO_4$

Place three to four drops of saturated ammonium molybdate into each of two test tubes and add concentrated hydrochloric acid dropwise until a white crystalline solid: molybdic acid precipitates. Let the solution settle, remove the supernatant liquid from both test tubes using a pipette, and save the product for the next experiment:

$$(NH_4)_2MoO_4(\mathbf{4d}) + \text{Conc. } H^+(\mathbf{dw}) \rightarrow MoO_3 \cdot H_2O$$

Molybdic acid, $H_2MoO_4$, is most appropriately described as hydrated molybdenum trioxide, $MoO_3{\cdot}H_2O$. Under certain conditions, a dihydrate can also be isolated.

**Note:** Semiconcentrated HCl, $H_2SO_4$, or $HNO_3$ can be used in the preparation of molybdic acid.

## 1.2 Amphoteric property of molybdic acid

Take two test tubes containing molybdic acid prepared at Experiment 1.1, then add dropwise 2N, NaOH into one of the test tubes and 2N, $H_2SO_4$ into the other and notice the color change. Save molybdyl sulfate solution to the next experiment:

a) $MoO_3 \cdot H_2O + 2H^+ \rightarrow MoO_2^{2+}(\text{molybdenyl}) + 2H_2O$

b) $MoO_3 \cdot H_2O + 2OH^- \rightarrow MoO_4^{2+} \cdot 2H_2O$

## 1.3 Preparation of molybdenyl hexacyanoferrate(II)

In the molybdyl sulfate solution prepared at Experiment 1.2, add solution of potassium hexacyanoferrate(II), $K_4[Fe(CN)_6]$, dropwise. You will observe red-brown precipitate that confirms the formation of molybdenyl hexacyanoferrate(II):

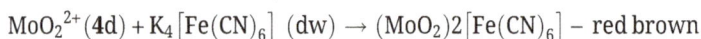

$$MoO_2^{2+}(\mathbf{4d}) + K_4\left[Fe(CN)_6\right]\ (\text{dw}) \rightarrow (MoO_2)2\left[Fe(CN)_6\right] - \text{red brown}$$

## 1.4 Preparation of sparingly soluble molybdates

Pour three to four drops of saturated ammonium molybdate solution into each of two test tubes and add the same volume of calcium chloride into the first and lead acetate into the other. Note the color of the corresponding molybdate precipitate:

$$Ca^{2+} + MoO_4^{2-} \rightarrow CaMoO_4$$

$$Pb^{2+} + MoO_4^{2-} \rightarrow PbMoO_4$$

## 1.5 Specific precipitation reaction for molybdate ion: identification as ammonium molybdatophosphate

Pour five to six drops of saturated ammonium molybdate solution into a test tube followed by addition of one to two drops of concentrated $HNO_3$. Add a drop of hydrogen phosphate solution and heat gently in water bath until you observe a yellow precipitate, which is characteristic of ammonium molybdatophosphate:

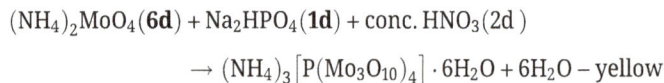

$$(NH_4)_2MoO_4(\mathbf{6d}) + Na_2HPO_4(\mathbf{1d}) + \text{conc. } HNO_3(2d)$$
$$\rightarrow (NH_4)_3\left[P(Mo_3O_{10})_4\right] \cdot 6H_2O + 6H_2O - \text{yellow}$$

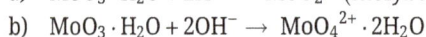

### 1.6 Preparation of thiomolybdate and Mo(VI) sulfide, $MoS_3$

Pour five to six drops of saturated ammonium molybdate solution into a test tube and then add ammonium sulfide dropwise. The mixture turns to red due to the formation of ammonium thiomolybdate.

Note: Ammonium sulfide has yellow color in solution due to the presence of ammonium polysulfide, $(NH_4)_2Sx$:

$$(NH_4)_2MoO_4(\mathbf{4d}) + (NH_4)_2S\,(\mathbf{dw}) \rightarrow MoS_4^{2-} + 8NH_3 + 4H_2O$$

**NB:** The red color is due to the presence of $MoS_4^{2-}$.

Now add dropwise 2N, HCl into the red precipitate or solution till it changes to black, which indicates the formation of molybdenum trisulfide:

$$MoS_4^{2-} + 2\text{ N, HCl (dw)} \rightarrow MoS_3 + H_2S$$

### 1.7 Peroxomolybdate

Place a few crystals of ammonium molybdate in a test tube and then add two to three drops of concentrated ammonia followed by two to three drops of 3% $H_2O_2$. Note the color change due to the formation of ammonium permolybdate, $MoOx^{2-}$. (X may vary from 5 to 8 depending on the concentration of $H_2O_2$ and the reaction temperature.) The equation for $MoO_5^{2-}$ and $MoO_6^{2-}$ are given below:

$$(NH_4)_2MoO_4(\mathbf{1\ g}) + \text{Conc. } NH_3(3d) + 3\,\%\ H_2O_2(3d) \rightarrow MoO_3(O_2)^{2-} + H_2O$$

$$(NH_4)_2MoO_4(\mathbf{1\ g}) + \text{Conc. } NH_3(3d) + 3\%\ 2H_2O_2(3d) \rightarrow MoO_2(O_2)_2^{2-} + H_2O$$

## 2 Preparation of lower valent molybdenum

### 2.1 Preparation of molybdenum blue, $Mo_2O_5$

Add three to four drops of saturated ammonium molybdate solution into a test tube and acidify it with two to three drops of 2N, HCl and then add stannous chloride solution in dropwise manner until the mixture turns blue and blue precipitate begins to settle:

$$(NH_4)_2MoO_4(\mathbf{4d}) + 2\text{ N, HCl (3d)} + SnCl_2\ (dw) \rightarrow Mo_2O_5 + Sn^{4+} + 3H_2O$$

$$MoS_4^{2-} + Sn^{2+} + 6H^+ \rightarrow Mo_2O_5 + Sn^{4+} + 3H_2O$$

### 2.2 Reduction of Mo(VI) to Mo(III)

Pour five to eight drops of saturated ammonium molybdate solution into a test tube and add one to two grains of zinc followed by three to four drops of concentrated HCl.

Note gradual the color change from blue to green and finally to brown, which is the characteristic property of Mo(III):

$$(NH_4)_2MoO_4(\mathbf{4d}) + \text{Conc. HCl}\,(4d) + Zn\,(2\,cs) \rightarrow \text{brown}$$

$$MoS_4{}^{2-} + 3Zn + 16H^+ \rightarrow Mo^{3+} + 3Zn^{2+} + 8H_2O$$

### 2.3 Identification of molybdenum by its hexathiocyanato Mo(III) complex

Add three drops of saturated ammonium molybdate solution into a test tube and two drops of 2N, HCl followed by potassium thiocyanate, KSCN, and two drops of stannous chloride solution. You will observe **red color** due to the formation of hexathiocyanato molybdenum(III) ion, $[Mo(SCN)_6]^{3-}$. Then add 5–10 drops of diethyl or petroleum ether and shake. After a minute, you will observe red ether layer. Thiocyanato complex is soluble in organic solvents:

$$(NH_4)_2MoO_4(\mathbf{3d}) + 2\text{ N HCl}\,(\mathbf{2d}) + KSCN(\mathbf{3d}) + SnCl_2(\mathbf{2d}) \rightarrow [Mo(SCN)_6]^{3-} + Sn^{4+} + 8H_2O$$

$$[Mo(SCN)_6]^{3-} + \text{ether} \Rightarrow \text{red ether layer}$$

$$2MoS_4{}^{2-} + 12SCN^- + Sn^{2+} + 16H^+ \rightarrow 2[Mo(SCN)_6]^{3-} + 3Sn^{4+} + 8H_2O$$

## 4.6 Observation

Color change
Evolution of gas
Sound of reaction

## 4.7 Result and discussion

Result: Write the balanced equation of the chemical reaction in ionic form.

Discussion: Discuss the outcomes of the experimental part and theoretical parts. Explain the errors if any.

## 4.8 Post lab questions

1. Describe the preparation and properties of $MoCl_5$.
2. Identify Mo(V) in aqueous solution by means of two precipitation and two redox reactions.
3. Complete the following equations:

$$MoO_3 + H_2 \longrightarrow$$

$$MoO_4^{2-} + S^{2-} + NH_4^{+} \longrightarrow$$

# Experiment 5

## 5.1 Chemistry of manganese

## 5.2 Objective

Students should be able to:
- prepare manganese(II) hydroxide and its oxidation by atmospheric oxygen,
- describe the preparation of metallic manganese (equations),
- understand the most common positive oxidation states of manganese in its compounds, and
- write down the equations for reactions in which $MnO_2$ acts as a reductant and an oxidant.

## 5.3 Theory

### The element

After iron, manganese is the most frequent heavy metal and occurs up to 0.1% in the Earth's crust. An important manganese ore is manganese dioxide, $MnO_2$, and others are rhodochrosite, $MnCo_3$, braunite, Mn(OH), and sulfides, MnS and $MnS_2$. A considerable amount of manganese is involved in certain iron, zinc, and cadmium minerals. It has one isotope: $^{55}Mn$ (100%).

### Preparation

Manganese-containing ores are roasted to Mn3O4 followed by reduction with aluminum:

$$3Mn_3O_4 + 8Al \rightarrow 9Mn + 4Al_2O_3$$

Pure metallic manganese is manufactured by electrolysis of Mn(II) salts, such as $MnSO_4$. Most of the manganese produced is processed as ferromanganese. This alloy can be obtained by the reduction of mixtures of Fe- and Mn-containing ores.

## 5.4 Chemicals and apparatus

| Chemicals | Apparatus |
|---|---|
| Solutions of $MnSO_4$ and $KMnO_4$ | Crucible |
| Solid of $KMnO_4$ and $MnO_2$ | Test tube |
| 2N, NaOH | Water bath |

https://doi.org/10.1515/9783111574349-006

(continued)

| Chemicals | Apparatus |
|---|---|
| 2N solution of ammonia | Bunsen burner |
| Distilled water | Test tube clamp |
| NH4Cl solution and concentrated HCl | |
| Sodium sulfide solution | |
| Bromine-water and potassium iodide solution | |
| 1% $CuSO_4$ | |
| 2N nitric acid | |

## 5.5 Procedure

### Preparation and reactions of manganese compounds

### 1 Manganese(II) compounds

#### 1.1 Preparation of manganese(II) hydroxide and its oxidation by atmospheric oxygen

Add three to four drops of manganese(II) salt solution to the test tube followed by two to three drops of 2N, NaOH. Pink $Mn(OH)_2$ is precipitated. Stir the mixture with a glass rod and note the color change to brown due to the oxidation of divalent manganese to the tetravalent state:

$$Mn^{2+} + 2OH^- \rightarrow Mn(OH)_2$$

$$2Mn(OH)_2 + O_2 \rightarrow 2MnO(OH)_2$$

Remember that the formation of higher positive oxidation state is, in general, facilitated under basic conditions.

#### 1.2 Action of ammonia on divalent manganese salts in the absence and presence of ammonium salts

In each of the test tubes, add three drops of 2N solution of ammonia. Add three drops of distilled water into one and the same volume of saturated $NH_4Cl$ solution into the other test tube. Now add three drops of $MnSO_4$ to each of the test tubes and note the occurrence of pink precipitate of $Mn(OH)_2$ in one of them. Write down the relevant equations to explain these results.

### 1.3 Preparation of manganese(II) sulfide and its oxidation by atmospheric oxygen

Place three to four drops of manganese(II) salt solution into the test tube and add two to four drops of ammonium or sodium sulfide solution. Note the precipitation of creamy pink MnS. Stir the mixture with glass rod and observe the color change to brown attributed to the formation of $MnO(OH)_2$:

$$Mn^{2+} + S^{2-} \rightarrow MnS$$

$$MnS + O_2 + H_2O \rightarrow MnO(OH)_2 + S$$

Pay attention to the last equation. It is an overall equation. Manganese(II) as well as sulfide are simultaneously oxidized by oxygen.

### 1.4 Oxidation of manganese(II) to its tetravalent state by bromine in alkaline medium

Add two drops of $MnSO_4$ solution to a test tube followed by three drops of 2N, NaOH. Manganese hydroxide is precipitated. Then add six drops of bromine water. The pink hydroxide turns immediately brown due to the formation of $MnO(OH)_2$:

$$MnO(OH)_2 + Br_2 + OH^- \rightarrow MnO(OH)_2 + H_2O + 2Br^-$$

Note that $Br_2$ in aqueous alkali forms hypobromite, $OBr^-$, which is the actual oxidizing agent.

### 1.5 Oxidation of Mn(II) to heptavalent manganese by bromine in alkaline solution with Cu(II) as a catalyst

Mn(II) is oxidized to Mn(VII) by hypobromite in the presence of Cu(II) salt acting as a catalyst. For the experiment, take two to three drops of $MnSO_4$ solution and add five drops of 1% $CuSO_4$ solution. In another test tube, prepare hypobromite by mixing five drops of bromine water with such an amount of 2N, NaOH as is necessary to decolorize the solution. Now add the hypobromite reagent to the Mn(II)/Cu(II) sample and boil for a minute. Let the precipitate settle and observe the violet color of the supernatant solution:

$$2Mn^{2+} + 5BrO^- + 6OH \rightarrow 2MnO_4^- + 5Br^- + 3H_2O$$

### 1.6 Oxidation of Mn(II) to heptavalent manganese by lead dioxide in acidic solution

Add a small quantity of lead dioxide to a test tube followed by six to eight drops of 2N nitric acid and then one drop of Mn(II) salt solution. Boil carefully. Then let the excess $PbO_2$ to settle and observe the color of the supernatant solution:

$$2Mn^{2+} + 5PbO_2 + 4H^+ \rightarrow 2MnO_4^- + 5Pb^{2+} + 2H_2O$$

## 2 Manganese(IV) compounds

### 2.1 Oxidizing properties of manganese dioxide

Place a micro-spatula full of manganese dioxide into a test tube and add two to three drops of concentrated HCl. Yellow chlorine gas evolves. If the reaction is not intense, heat the test tube gently over a small flame:

$$MnO_2 + 4HCl \rightarrow MnCl_2 + Cl_2 + 2H_2O$$

### 2.2 Reducing properties of manganese dioxide: identification of manganese by oxidative fusion

Place a small pellet of NaOH and the same amount of crystalline potassium nitrate into a crucible and heat in a clay triangle (or asbestos wire gauze) over a burner and melt the mixture. Continue heating and add a very small quantity of manganese dioxide. A green melt is obtained due to the formation of manganate(VI), $MnO_4^{2-}$. A shade of the blue may be attributed to manganate(V), $MnO_4^{3-}$, a side product of the reaction. Cool the melt and continue immediately with the following experiment:

$$3MnO_2 + 3M^IN O_3 + 4H^IOH \rightarrow 3M^I_2MnO_4 + 2NO + 2H_2O$$

## 3 Manganese(VI) compounds

### 3.1 Disproportionation of manganate(VI)

Add water to the melt prepared in the previous experiment (as soon as it is cold) and stir the crucible content with the glass rod. Transfer five to six drops of the resulting green solution into each of the two test tubes. Add two to three drops of 2N acetic acid in one test tube. The solution turns violet immediately and the brown precipitation settles. The solution in the other test tube is, however, fairly stable. Only after a longer time, it turns slowly violet with concomitant precipitation of brown solid.

The disproportionation of manganese(VI) in water can be described by the following equation:

$$3MnO_4^{2-} + 2H_2O \rightarrow 2MnO_4^- + MnO_2 + 4OH^-$$

Thus, valence disproportionation may be defined as a redox process with a species in a medium oxidation state being transferred into two other species, one in a higher and another in a lower oxidation state. Manganate(VI) in neutral or acidic solution undergoes such a reaction.

### 3.2 Oxidizing properties of manganate(VI): reaction with sodium sulfite

Place small quantities of crystalline sodium sulfite into test tube containing 5–10 drops of green manganate(VI) solution. The solution decolorizes and a brown precipitate settles. The latter is $MnO_2$:

$$MnO_4^{2-} + SO_3^{2-} + 2H_2O \rightarrow MnO_2 + SO_4^{2-} + 2OH^-$$

## 4 Manganese(VI) compounds

### 4.1 Preparation and properties of permanganic hydride

This experiment is to be performed only by the instructor. Permanganic hydride is extremely explosive. Manganese(VII) oxide, $Mn_2O_7$, is a very powerful oxidizing agent. Oxidation of organic matter usually proceeds under ignition or explosion. $Mn_2O_7$ proves to be thermodynamically unstable with respect to decomposition into $MnO_2$ and $O_2$. Therefore, it explodes on heating. For the instructor: Drop a few crystals of potassium permanganate into a test tube and add some drops of concentrated sulfuric acid. The mixture warms up and attains a greenish dark brown color. Manganese heptaoxide, a dark brown liquid, has been formed. Now take a piece of filter paper and fold it into a paper cone. Put it on an asbestos wire gauze in a hood. Then add a few drops of the $Mn_2O_7$-containing liquid onto the top of the paper cone. The strong oxidizing agent turns the paper black. The heat of reaction will ignite the paper finally:

$$2KMnO_4 + H_2SO_4 \rightarrow Mn_2O_7 + K_2SO_4 + H_2O$$

### 4.2 Thermal decomposition of potassium permanganate

Place three to four crystals of potassium permanganate into a test tube, clamp it on a stand in the horizontal position, and heat over a low flame to decompose the substance completely to $K_2MnO_4$, $MnO_2$, and $O_2$. Cool the test tube and add five to six drops of distilled water to the dry residue. Observe a blue solution and a brown precipitate:

$$2KMnO_4 + H_2SO_4 \rightarrow K_2MnO_4 + MnO_2 + O_2$$

The evolution of oxygen and the completeness of the decomposition should be checked with a glowing splint. If the oxygen test fails, try it again with some more $KMnO_4$.

### 4.3 pH dependence of the oxidizing properties of potassium permanganate: reaction with sodium sulfite in acidic, neutral, and alkaline mediaum

The oxidation potential of permanganate in aqueous solution depends on the pH value. In acidic medium, $MnO_4^-$ is reduced to $Mn^{2+}$, in neutral and slightly alkaline systems $MnO_2$ precipitates, and in strongly basic solution green manganate(VI) is formed.

Place three to four drops of potassium permanganate solution into each of three test tubes. Add two drops of 2N sulfuric acid into one test tube, two drops of water into the second test tube, and two to three drops of 2N aqueous NaOH into the third one. Take a concentrated solution of sodium sulfite and add it dropwise to the three samples. Record carefully your observations:

$$2MnO_4^- + 5HSO_3^{3-} + 5H^+ \xrightarrow{\text{acidic}} 2Mn^{2+} + 5SO_4^{2-} + 3H_2O$$

$$2MnO_4^- + 3SO_3^{2-} + H_2O \xrightarrow{\text{sl - alkali}} 2MnO_2 + 3SO_4^{2-} + 2OH^-$$

$$2MnO_4^- + SO_3^{2-} + 2OH^- \xrightarrow{\text{str - alkali}} 2MnO_4^{2-} + SO_4^{2-} + H_2O$$

### 4.4 Reaction of potassium permanganate with potassium iodide in acidic, neutral, and alkaline media

Add three to four drops of potassium permanganate solution into each of three test tubes. Add two drops of 2N sulfuric acid into one test tube, two drops of water into the second test tube, and three to four drops of 2N aqueous NaOH into the third one. In acidic solution, dropwise addition of potassium iodide solution first decolorizes the system under precipitation of black solid iodide. With excess of iodide, the iodine is redissolved to give a brown solution containing the triiodide anion, $I_3^-$:

$$MnO_4^- + 10I^- + 16H^+ \rightarrow 2Mn^{2+} + 5I_2 + 8H_2O$$

In neutral $MnO_4^-$ solution, KI forms the brown $MnO_2$ precipitate, whereas in alkaline medium the color of the solution changes to green. Interpret the result and write ionic equations. Note that in the presence of alkali, iodate ($IO_3^-$) is formed rather than iodine.

### 4.5 Oxidation of ferrous sulfate by potassium permanganate

Add three to four drops of potassium permanganate to a test tube. Then add three to four drops of 2N sulfuric acid and few crystals of Mohr's salt $(NH_4)_2Fe(SO_4)_2$. The color of $MnO_4^-$ will disappear and pale yellow color will form due to the formation of $Fe^{3+}$:

$$MnO_4^- + 5Fe^{2+} + 8H^+ \rightarrow 2Mn^{2+} + 5Fe^{3+} + 4H_2O$$

### 4.6 Oxidation of hydrogen peroxide by potassium permanganate

Add three to five drops of potassium permanganate into a test tube followed by two to three drops of 2N sulfuric acid, and then add three to four drops of 10% hydrogen peroxide. The color of $MnO_4^-$ will disappear and oxygen is formed:

$$MnO_4^- + 5H_2O_2 + 6H^+ \rightarrow 2Mn^{2+} + 5O_2 + 8H_2O$$

### 4.7 Oxidation of alcohol by potassium permanganate

Add three to four drops of potassium permanganate to two test tubes followed by two drops of 2N sulfuric acid in one of the test tubes and equal amount of 2N, NaOH into the other. Then add three drops of alcohol in both test tubes. Heat gently the acidic solution over the small flame, and the alcohol reduces from $MnO_4^-$ to $Mn^{2+}$. However, in an alkaline medium, the color changes to green and eventually to brown, which is the characteristic color of $MnO_2$:

$$MnO_4^- + 5C_2H_5OH + 6H^+ \rightarrow 2Mn^{2+} + 5CH_3CHO + 8H_2O$$

$$MnO_4^- + 5C_2H_5OH + 2OH^- \rightarrow 2MnO_4^{2-} + 5CH_3CHO + 8H_2O$$

$$2MnO_4^- + 5C_2H_5OH \rightarrow MnO_2 + 5CH_3CHO + 2OH^-$$

### 4.8 Synproportionation of manganese(II) and manganese(VII)

Place three to four drops of potassium permanganate into two test tubes and equal amount of manganese(II) sulfate. The color of $MnO_4^-$ disappears and brown precipitate is formed, which indicates the formation of $MnO_2$. Check the pH of the solution using the litmus paper:

$$MnO_4^- + Mn^{2+} + 2H_2O \rightarrow MnO_2 + 4H$$

## 5.6 Observation

Color change
Evolution of gas
Sound of reaction

## 5.7 Results and discussion

## 5.8 Post lab questions

1. Compile all the oxides and hydroxides or acids of manganese in various oxidation states and discuss their acidic and basic behavior.
2. The following substances are added to manganese(II) sulfate dissolved in water in open air:
   a. Potassium hydroxide
   b. Ammonium sulfide
   c. Potassium permanganate. Write the equations.

3. How can the following substances be prepared from $MnSO_4$?
   A. Mn $(OH)_2$; B. $MnO_2$; C. $KMnO_4$. Write down the equations.
4. Write the equations for reactions in which $MnO_2$ acts as
   a. reductant and b. an oxidant

# Experiment 6

## 6.1 Chemistry of iron

## 6.2 Objective

Students should be able to:
- describe the preparation of metallic iron in industry and laboratory scales,
- characterize the solubility of iron in acids,
- characterize the main types of ions and compounds in these oxidation states, and
- know how a ferrous salt can be converted into ferric salt and how this process can be reversed.

## 6.3 Theory

### Occurrence

In the frequency chart of elements, iron holds the fourth position by 5.1%. After aluminum, it is the most frequently occurring metal. Metallic iron is the main constituent of many meteorites.

Iron is a heavy metal and occurs in various ores (magnetite, $Fe_3O_4$, hematite, $Fe_2O_3$, brown iron stone, $Fe_2O_3 \cdot XH_2O$), sulfidic ores (pyrite, $FeS_2$; magnetic pyrites, $Fe_{1-x}S_x$), and spathic iron ore, $FeCO_3$.

Iron exists as the following isotopes: $^{54}Fe$(5.84%), $^{56}Fe$(91.66%), $^{57}Fe$(2.17%), and $^{58}Fe$(0.33%).

### Preparation

The enormous cultural importance of iron originates from its great strength and the ability of forming alloys with carbon and many metals. All various iron ores are first of all upgraded to iron oxide. The latter is reduced in blast furnace by means of carbon and CO. In addition, coke serves as a heating medium and forms compounds with iron that modifies the properties of the final product. Mineral impurities are converted into slag.

The reaction taking place in the blast furnace is of complex nature. At higher temperatures, the Boudouard equilibrium is of importance:

$$CO_2 + C \rightarrow 2CO$$

In the upper part of the furnace, CO acts as a reducing agent because below 7,000 °C, CO proves to have a higher reducing power than carbon. In the lower part of the oven, elementary carbon comes directly into action as a reducing agent:

https://doi.org/10.1515/9783111574349-007

$$FeO + CO \rightarrow Fe + CO_2$$

$$FeO + C \rightarrow Fe + CO$$

Above 10,000 °C, carbon may react with iron to give $Fe_3C$.

Chemically pure iron is obtained by thermolysis of $Fe(CO)_5$:

$$Fe(CO)_5 \xrightarrow{250\,°C} Fe + 5CO\ 250\,°C$$

Other methods are the reduction of iron oxide prepared from iron(II) oxalate, carbonate, or nitrate by hydrogen at 7,000 °C. Iron powder prepared in this way is usually pyrophoric.

## 6.4 Chemicals and apparatus

| Chemicals | Apparatus |
|---|---|
| Iron fillings | Test tube |
| Aqueous solutions of Mohr's salts $(NH_4)_2Fe(SO_4)_2{\cdot}6H_2O$ | Spatula |
| Iron(III) chloride, $FeCl_3{\cdot}6H_2O$ | Bunsen burner |
| Iron(II) oxalate ($Fe(C_2O_4)_2$) | |
| 2N, HCl, $H_2SO_4$, $HNO_3$ | |
| Conc. $H_2SO_4$ | |
| Potassium or ammonium thiocyanate | |
| 2N, NaOH and 3% $H_2O_2$ | |
| Tin(II) chloride, mercuric nitrate | |
| Copper(II) sulfate | |
| Ferric chloride salt solution | |
| $Na_2CO_3$ solution | |

## 6.5 Procedure

### 1 Preparation and properties of pyrophoric iron

#### 1.1 Preparation and properties of pyrophoric iron

Place one spatula full of iron(II) oxalate, $Fe(C_2O_4)_2$. Clamp the tube in a horizontal position with an iron stand and heat it on the Bunsen burner gently until the yellow color turns to black and gas evolution stops. Remove the tube from the burner and stopper it immediately. After cooling, pour the product on a paper from the height of 25–30 cm above the paper. You will observe the spark of burning iron:

$$Fe(C_2O_4)_2 \xrightarrow{heat} FeO + CO + CO_2$$

$$FeO \xrightarrow{heat} Fe + Fe_3O_4$$

Iron oxidizes by air to $Fe_3O_4$:

$$Fe + 2O_2 \rightarrow Fe_3O_4$$

### 1.2 Reaction of Fe with acids

Add five drops of 2N, HCl, $H_2SO_4$, $HNO_3$ and three drops of conc. $H_2SO_4$ to four test tubes separately. Add one micro-spatula full of iron filling to each test tube and observe the reaction. Then add a drop of potassium or ammonium thiocyanate tubes containing dilute acids, and the appearance of red color indicates the presence of iron (III). Note that iron(III) thiocyanate is unstable so the red color persists only for 1 or 2 min. Heat the tube containing conc. $H_2SO_4$ and iron filling. After heating, transfer two to three drops to another test tube and dilute with 1 mL distilled water. Then add thiocyanate reagent and observe the reaction. Write the equation.

### 1.3 Displacement of some metals from their salts by iron

Add 10 drops of tin(II) chloride, mercuric nitrate, and copper(II) sulfate to three test tubes separately. Then put cleaned iron nails into each test tube and interpret your observation.

## 2 Iron(II) compounds

Due to the partial oxidation in air, ferrous salts always contains traces of Fe(III). The most stable crystalline Fe(II) is Mohr's salt $(NH_4)_2Fe(SO_4)_2{\cdot}6H_2O$; a double salt dissociates completely in water.

### 2.1 Preparation of ferrous hydroxide and it oxidation by atmospheric oxygen

Place three to four drops of Mohr's salt into a test tube and add 2N, NaOH until white flakes of $Fe(OH)_2$ are precipitate.

Note: If the solution contains trace of $Fe^{3+}$, the color of the precipitate becomes pale green:

$$Fe^{2+} + 2OH^- \rightarrow Fe(OH)_2$$

$$4Fe(OH)_2 + O_2 + 2H_2O_4 \rightarrow Fe(OH)_3$$

### 2.2 Basic character of ferrous hydroxide

Place ferrous hydroxide, $Fe(OH)_2$, into two test tubes and add several drops of 2N, $H_2SO_4$ in one of the tubes and 2N, NaOH into the other. $Fe(OH)_2$ dissolves in acid:

$$4Fe(OH)_2 + 2H^+ \rightarrow 4Fe^{2+} + 2H_2O$$

### 2.3 Preparation of sparingly soluble ferrous carbonate

Add five drops of distilled water to a test tube and boil it to remove dissolved oxygen. Dissolve a few crystals of Mohr's salt in this oxygen-free water and add three to five drops of $Na_2CO_3$ solution.

You will observe white precipitate that indicates ferrous carbonate:

$$Fe^{2+} + 2CO_3^{2-} \rightarrow FeCO_3$$

### 2.4 Preparation of ferrous sulfide

Add five to six drops of Mohr's salt to each of two test tubes followed by a drop of freshly prepared $H_2S$ to one test tube and $(NH_4)_2S$ to the other tube. Black FeS precipitates only in slightly alkaline medium. Explain the influence on the solubility and check the solubility of FeS in dilute HCl.

### 2.5 Reaction of iron(II) with potassium hexacyanoferrate(III) – Turnbull's blue

Add two to three drops of Mohr's salt to a test tube and then a drop of potassium hexacyanoferrate(III), $K_3[Fe(CN)_6]$. A deep blue precipitate of Turnbull's blue will appear:

$$Fe^{2+} + 2\left[Fe(CN)_6\right]^{3-} \rightarrow K_3\left[Fe(CN)_6\right]^2$$

**Note:** This test is sensitive to $Fe^{2+}$.

### 2.6 Reducing property of divalent iron

#### 2.6.1 Reduction of nitric acid, $HNO_3$

Add five drops of fresh Mohr's salt to each of two test tubes, then a drop of concentrated $HNO_3$ to one test tube and heat until gas evolution stops, finally cool it down. The color turns from green to yellow. The intermediate brown color is due to the formation of unstable $[Fe(NO)(H_2O)_5]^{2+}$. This is the identification of nitrate (ring test) and nitrite:

$$Fe^{2+} + NO_3^- + 4H^+ \rightarrow Fe^{3+} + NO + 2H_2O$$

#### 2.6.2 Reduction of potassium permanganate, $MnO_4^-$

Add five to six drops of $MnO_4^-$ to a test tube followed by two drops of concentrated $H_3SO_4$. Then add Mohr's salt solution dropwise until you observe decolorization of the mixture, which infers the reduction of $MnO_4^-$ to $Mn^{2+}$. Write the equation.

### 2.6.3 Reduction of $H_2O_2$

Add five to six drops of Mohr's salt to each of two test tubes and to one sample add two to three drops of 2N, $H_2SO_4$ followed by the same volume of 3% $H_2O_2$. Observe the color change from green to yellow. Check for $Fe^{3+}$ in both test tubes using ammonium thiocyanate reagents:

$$Fe^{2+} + H_2O_2 + 2H^+ \rightarrow Fe^{3+} + 2H_2O$$

The evolution of oxygen is the side product from disproportionation reaction of $H_2O_2$ because of the transition metal catalyst.

## 3 Iron(III) compounds

### 3.1 Detection of iron(III) as a constituent of the decomposition products of the iron oxalate thermolysis

Take 1 g of iron(II) oxalate in a crucible fixed in clay triangle and heat over a burner until the color change occurs. Cool the resulting residue and dissolve a small quantity of sample in a test tube with 2N, HCl. Test for $Fe^{3+}$. Write the equation for the reaction.

### 3.2 Preparation and properties of ferric hydroxide

Take five to six drops of ferric chloride salt solution in two test tubes and add three to four drops of 2N, NaOH, you will observe brown precipitate due to the formation of $Fe(OH)_3$.

Note: If the solution contains traces of $Fe^{3+}$, the color of the precipitate becomes pale green:

$$Fe^{3+} + 3OH^- \rightarrow Fe(OH)_3$$

Then add 2N, $H_2SO_4$ into one of the tubes containing $Fe(OH)_3$ and 2N, NaOH into the other and observe changes. Compare $Fe(OH)_3$ with $Cr(OH)_3$, $Mn(OH)_2$, and $Fe(OH)_2$ using table.

**Note:** $Fe(OH)_3$ is insoluble in NaOH.

### 3.3 Specific reactions of $Fe^{3+}$

### 3.3.1 Reaction with potassium thiocyanate

Take five to six drops of ferric chloride salt solution in a test tube and add a drop of potassium thiocyanate, KSCN, you will observe red which is the characteristic of $Fe(SCN)_3$:

$$Fe^{3+} + 3SCN^- \rightarrow Fe(SCN)_3$$

Repeat the experiment using freshly prepared Mohr's salt.

### 3.3.2 Reaction with potassium hexacyanoferrate(II), $K_4[Fe(CN)_6]$

Take five to six drops of ferric chloride salt solution in a test tube and add a drop of $K_4[Fe(CN)_6]$, you will observe deep blue precipitate. Prussian blue indicates the presence of Fe(III):

$$Fe^{3+} + 3\left[Fe(CN)_6\right]^{4-} \rightarrow Fe_4\left[Fe(CN)_6\right]^3$$

**Note:** This test is extremely sensitive to $Fe^{3+}$.

## 3.4 Oxidizing properties of iron(III)

### 3.4.1 Oxidation of sodium sulfite, $Na_2SO_3$

Take five to six drops of ferric chloride salt solution in a test tube and add a few crystals of $Na_2SO_3$. At first you will observe red color due to the formation of Fe(III) sulfite complex but fades quickly and change to pale green, indicating the presence of $Fe^{2+}$. Check it by adding $K_3[Fe(CN)_6]$.

### 3.4.2 Oxidation of hydrogen sulfide, $H_2S$

Take three to four drops of ferric chloride salt solution in a test tube and add two to three drops of freshly prepared hydrogen sulfide solution. The color of $Fe^{3+}$ fades and the colloidal precipitate of sulfur turns the mixture white and turbid:

$$Fe^{3+} + H_2S \rightarrow Fe^{2+} + S + 2H^+$$

## 3.5 Preparation of ferric sulfide

Take three to four drops of ferric chloride salt solution in a test tube and add two to three drops of ammonium sulfide solution. Black precipitate forms but fades soon due to its instability and decomposes to black FeS and white S:

$$Fe^{3+} + 3S_2^{-} \rightarrow Fe_2S_3$$

$$Fe_2S_2 \rightarrow 2FeS + S$$

To check the presence of S, add 2N, $H_2SO_4$; the acid dissolves FeS but leaves S.

# 6.6 Observation

Color change
Evolution of gas
Sound of reaction

## 6.7 Results and discussion

## 6.8 Review problems

1. Characterize the solubility of iron in acids.
2. What is understood by the term “rusting”? How can it be prevented?

# Experiment 7

## 7.1 Chemistry of cobalt

## 7.2 Objective

Students should be able to:
- discuss the preparation of metallic cobalt,
- describe the reaction of cobalt compound with alkali, with ammonia and carbonates,
- observe the oxidation of $Co(OH)_2$ with hydrogen peroxide, and
- identify Co(II) as hexanitrocobaltate(III).

## 7.3 Theory

### Occurrence

The content of cobalt in the Earth's crust amounts to 0.001%. Cobalt is closely associated with nickel, mainly in the form of arsenic and sulfur compounds. The most important minerals are skutterudite, $(Co,Ni)As_3$, cobaltite, CoAsS, and smalite, $CoAs_2$. The metallic element is itself found in meteorites with 0.5–2.5% cobalt.

### Preparation

The combination of chemical, thermal, and electrolytic methods of preparation for cobalt depends on the nature of the minerals available. Initially, all Ni–Co–Cu ores are roasted and upgraded to give oxides, sulfates, and arsenates. The latter are then dissolved in sulfuric acid and chemically separated. Cu, Pb, Bi, and others are precipitated by $H_2S$, As, and Fe by means of limestone, and Co by chlorinated lime as $Co(OH)_3$. Nickel remains in solution. $Co(OH)_3$ is calcined followed by reduction with charcoal or hydrogen:

$$2Co(OH)_3 \rightarrow Co_2O_3 + H_2O$$

$$Co_2O_3 + 3H_2 \rightarrow 2Co + 3H_2O$$

$$Co_2O_3 + 3C \rightarrow 2Co + 3CO$$

Pure cobalt may be obtained by $H_2$ reduction of recrystallized cobalt complexes for example, $[Co(NH_3)_5Cl]Cl_2$.

https://doi.org/10.1515/9783111574349-008

## 7.4 Chemicals and apparatus

| Chemicals | Apparatus |
|---|---|
| - Solution of $Co(NO_3)2.6H_2O$ or $CoCl_2 \cdot 6H_2O$ | Test tube |
| - Concentrated aqueous solution of NaOH | Water bath |
| - 2N, $H_2SO_4$ and 2N, NaOH | Bunsen burner |
| - Dilute ammonia solution | |
| - $NH_4Cl$ and 6% $H_2O_2$ | |
| - Sodium carbonate | |
| - $NH_4S$, bromine water, and KCN | |

## 7.5 Procedure

### 1 Preparation and reactions of cobalt compounds

#### 1.1 Reaction of Co(II) with alkali hydroxide

Take five drops of Co(II) salt solution in two test tubes and slowly add two to three drops of concentrated aqueous solution of NaOH to it. Blue precipitate will be observed and the precipitate will turn to pink through heating:

$$Co^{2+} + 2OH^- \rightarrow Co(OH)_2$$

#### 1.2 Solubility of Co(II) hydroxide, $Co(OH)_2$

Take seven to eight drops of $Co(OH)_2$ in each of two test tubes and two to three drops of 2N, $H_2SO_4$ in one test tube and equal volume of 2N, NaOH in the other. $Co(OH)_2$ dissolves readily in acid but partly in NaOH by forming a deep blue complex, $[Co(OH)_4]^{2-}$. Write the equations.

#### 1.3 Reaction of Co(II) with ammonia

Take five drops of Co(II) salt solution in two test tubes and add dilute ammonia solution to it in a dropwise manner. Blue precipitate is formed but turns to pink in air due to oxidation and forms Co(III) ammine complexes:

$$Co^{2+} + O_2 + 6H_2O + 20\,NH_3 \rightarrow \left[Co(NH_3)_5(H_2O)\right]^{3+} + 4\,OH^-$$

Repeat the above experiment in the presence of $NH_4Cl$; no precipitate will be formed but yellowish solution changes to pink.

### 1.4 Reaction of Co(II) with carbonate, $CO_3^{2-}$

Take five drops of Co(II) salt solution in a test tube and add an excess of sodium carbonate solution. Depending on the concentration, bluish or reddish precipitate of basic carbonate is formed.

### 1.5 Reaction of Co(II) with ammonium sulfide, $NH_4S$

Take five drops of Co(II) salt solution in a test tube and add a few drops of $NH_4S$ to form black CoS precipitate. Repeat the experiment using gaseous $H_2S$ or $H_2S$–water. CoS is soluble in $HNO_3$ but hardly soluble in HCl:

$$Co^{2+} + S_2^{-} \rightarrow CoS$$

$$3CoS + HNO_3 + 6H^{+} \rightarrow Co^{2+} + 3S + 2NO + 4H_2O$$

### 1.6 Oxidation of $Co(OH)_2$ with hydrogen peroxide

Add seven to eight drops of $Co(OH)_2$ to a test tube followed by three drops of 6% $H_2O_2$ and boil. Observe the formation of black solid, $Co(OH)_3$:

$$Co(OH)_2 + H_2O_2 \rightarrow Co(OH)_3$$

### 1.7 Oxidation of $Co(OH)_2$ with bromine water

Add seven to eight drops of $Co(OH)_2$ to a test tube followed by two to three drops of bromine water and observe the formation of black precipitate, $Co(OH)_3$. Write the equation.

### 1.8 Reaction of Co(II) with potassium cyanide, KCN

Take five drops of Co(II) salt solution in a test tube and add KCN in a dropwise manner. Red brown $Co(CN)_2$ precipitate is formed. In excess KCN, $Co(CN)_2$ dissolves and forms $[Co(CN)_6]^{4-}$. Then add two to three drops of 6% $H_2O_2$ color of the solution that turns to blue due to the formation of complex anion, $[Co(CN)_6]^{3-}$:

$$[Co(CN)_6]^{4-} + H_2O_2 \rightarrow [Co(CN)_6]^{3-} + 2OH^{-}$$

### 1.9 Reaction of Co(II) with thiocyanate

Take five drops of Co(II) salt solution in a test tube and add the same volume of concentrated ammonium or potassium thiocyanate solution. Then cover the mixture with 1 mL of ether. In neutral solution, blue $Co(SCN)_2$ is formed, which is soluble in organic solvent.

Repeat the experiment by adding two drops of dilute acetic acid, and the same observation will appear even if the result is a complex acid, $H_2[Co(SCN)_4]$. It dissolves in ether layer:

$$Co^{2+} + 2\,SCN^- \rightarrow Co(SCN)_2$$

$$Co^{2+} + 4SCN^- + 2H^+ \rightarrow H_2\left[Co(SCN)_4\right]$$

## 7.6 Observation

Color change
Evolution of gas and sound of reaction

## 7.7 Review questions

1. Name important cobalt-containing minerals. To which class of compounds do they belong?
2. Discuss the preparation of metallic cobalt.
3. Which are the most common ONs of cobalt in its compounds? Discuss their stability.
4. $Co(NO_3)_2$ is stable in aqueous solution but $Co(NO_3)_3$ is not. However, $Co(NO_3)_2$ is rapidly oxidized in the presence of cyanide. Explain these phenomena.
5. Describe the oxides and hydroxides of cobalt and account for their behavior with respect to acids and bases, respectively.
6. Compile analytically significant precipitation, redox, and complex formation reactions of cobalt.

# Experiment 8

## 8.1 Chemistry of nickel

## 8.2 Objective

Students should be able to:
- discuss the methods of preparation for crude and pure nickel,
- describe the so-called Mond process,
- understand how nickel reacts with aqueous HCl and $HNO_3$, and
- discuss the reaction of Ni(II) with cyanide, thiocyanate, and dimethylglyoxime.

## 8.3 Theory

### Occurrence

With 0.016% nickel contained in the Earth's crust, this element occurs with a higher frequency than Cu, Zn, and Pb all together. However, its industrial preparation is much more difficult because nickel is widely spread in nature and larger deposits are rare.

Nickel ores occur as sulfides, oxides, silicates, and amides. Elementary nickel has been found in meteorites (e.g., 91% Fe, 8% Ni, and 0.6% Co).

Nickel's isotopes are $^{58}Ni$(67.76%), $^{60}Ni$(26.26%), $^{61}Ni$(1.25%), $^{62}Ni$(3.66%), and $^{64}Ni$(1.16%).

### Preparation

The metallurgy of nickel is rather complicated. In principle, the ores are upgraded, concentrated, transferred into nickel sulfide, oxidized to NiO by roasting with hot air, and eventually reduced by charcoal to give crude nickel.

Pure nickel is prepared by the Mond process. The crude metal reacts at 50 °C with carbon monoxide yielding tetracarbonyl nickel(0), $Ni(CO)_4$, a poisonous, colorless liquid with b.p. 43 °C. $Ni(CO)_4$ decomposes again at 2,000 °C to give very pure nickel (99.99%):

$$\mathrm{Ni} + 4\mathrm{CO} \xrightarrow{50\,^{\circ}\mathrm{C}} \mathrm{Ni(CO)_4}$$

Other methods of purification are electrolytic refining and zone melting.

https://doi.org/10.1515/9783111574349-009

## 8.4 Chemicals and apparatus

| Chemicals | Apparatus |
|---|---|
| – Solution of $NiCl_2{\cdot}6H_2O$ | Test tube |
| – 2N, NaOH | Bunsen burner |
| – Crystal of $K_2S_2O_8$ | Water bath |
| – Chlorine water and KI solution | |
| – 2N, HCl and ammonia | |
| – Saturated $Na_2CO_3$ | |
| – Ammonium sulfide | |
| – Dilute sodium cyanide | |

## 8.5 Procedure

### 1 Reaction of Ni(II) with alkali hydroxide

Take five drops of Ni(II) salt solution in a test tube and add dropwise 2N, NaOH. You will observe green precipitate of $Ni(OH)_2$, which is insoluble in excess NaOH. Save the precipitate for the following experiments:

$$Ni^{2+} + 2OH^- \rightarrow Ni(OH)_2$$

### 2 Oxidation of Ni(II) hydroxide by potassium peroxodisulfate, $K_2S_2O_8$

Take $Ni(OH)_2$ (from Experiment 1) in a test tube and add crystals of $K_2S_2O_8$, then you will observe blue hydrous Ni(II) oxide:

$$Ni(OH)_2 + S_2O_8{}^{2-} + H_2O \rightarrow NiO(OH) + 2SO_4{}^{2-} + 2OH$$

**Note:** $Ni(OH)_2$ is not oxidized by $H_2O_2$.

### 3 Oxidation of Ni(II) hydroxide by chlorine

To $Ni(OH)_2$ precipitate, add two to three drops of chlorine water and you will observe black precipitate, which indicates the oxidation of $Ni(OH)_2$ to NiO(OH). Write the equation. When the black precipitate settles, decant the supernatant solution and add three to four drops of KI solution followed by two drops of 2N, HCl. The black precipitate dissolves and Ni(III) reduces to Ni(II) eventually turns to brown due to the formation of triiodide ion:

$$NiO(OH) + I^- + 3H^+ \rightarrow Ni^{2+} + \frac{1}{2}I_2 + 2H_2O$$

### 4 Reaction of Ni(II) with aqueous ammonia

Take 0.5 mL of Ni(II) salt solution in a test tube and add ammonia dropwise. At first, green $Ni(OH)_2$ precipitates but it dissolves in excess ammonia through forming blue hexamine complex $[Ni(NH_3)_6]^{2+}$. Write down the equation.

### 5 Reaction of Ni(II) with sodium carbonate

Take five drops of Ni(II) salt solution in a test tube and add several drops of saturated $Na_2CO_3$. The green precipitate confirms the formation of basic carbonate.

Note: The protolysis reaction of carbonate with water increases pH and initiates partial precipitation of $Ni(OH)_2$ mixed with $NiCO_3$. Write the equation.

### 6 Reaction of Ni(II) with sulfide

Take five drops of Ni(II) salt solution in a test tube and add several drops of ammonium sulfide. Black NiS will be precipitated:

$$Ni^{2+} + S^{2-} \rightarrow NiS$$

**Note:**
- NiS is formed only in neutral or slightly alkaline medium.
- NiS is soluble in mineral acids, immediately oxidizes to Ni(OH)S and $Ni_2S_3$ like CoS, and dissolves in $HNO_3$ and $HAc/H_2O_2$.

### 7 Reaction of Ni(II) with sodium cyanide

Take five drops of Ni(II) salt solution in a test tube and add dilute sodium cyanide solution in dropwise manner. The light green precipitate indicates the formation of $Ni(CN)_2$, which dissolves in excess $CN^-$ yielding $[Ni(CN)_4]^{2-}$ yellow solution:

$$Ni^{2+} + 2CN^- \rightarrow Ni(CN)_2$$

$$Ni(CN)_2 + 2CN^- \rightarrow [Ni(CN)_4]^{2-}$$

### 8 Reaction of tetracyanonickelate(II) $[Ni(CN)_4]^{2-}$ with NaOH/$Br_2$

***Attention!!! do this experiment in a hood***

Dissolve $Ni(CN)_2$ in a limited amount of $CN^-$ and add three drops of 2N, NaOH followed by three to four drops of bromine water. You will observe the black color:

$$[Ni(CN)_4]^{2-} + 6OH^- + 9Br_2 \rightarrow 2NiO(OH) + 10Br^- + 8BrCN + 2H_2O$$

**NB:** Avoid excess $CN^-$ because it reacts first with $Br_2$:

$$CN^- + Br_2 \rightarrow BrCN + Br^-$$

### 9 Complex formation of Ni(II) with pyridine

To 1 mL of Ni(II) salt solution, add 1 mL of aqueous potassium thiocyanate and a few drops of pyridine. A pale green, sparingly soluble complex salt $[Ni(py)_6](SCN)_2$ is precipitated. Write down the equation.

## 8.6 Observation

Color change
Evolution of gas
Sound of reaction

## 8.7 Results and discussion

## 8.8 Post lab questions

1. Discuss the methods of preparation for crude and pure nickel.
2. Describe the so-called Mond process.
3. How does nickel react with aqueous HCl and $HNO_3$?
4. Discuss the reaction of Ni(II) with cyanide, thiocyanate, and dimethylglyoxime.
5. Find some simple reactions to distinguish between Ni(II), Co(II), Fe(II), and Mn(II).
6. How is it possible to stabilize high and low oxidation states of an element? Take nickel as an example.

# Experiment 9

## 9.1 Chemistry of copper

## 9.2 Objective

Students should be able to:

- prepare copper from copper(II) oxide,
- understand why it is possible to use anhydrous $CuSO_4$ as an indicator for traces of moisture,
- observe the reaction of copper(II) oxide with dilute acids,
- understand the reducing properties of metallic copper,
- distinguish between Cu(I) and Cu(II) compounds by making uses of their magnetic properties, and
- propose at least three confirmatory tests for Cu(II) in aqueous solution.

## 9.3 Theory

### Occurrence

The Earth's crust contains about 100 g/t copper. This element occupies the 26th position in the frequency list and is, therefore, a rather common metal. Most copper ores are sulfides (more than 90%) with chalcopyrite, $CuFeS_2$, at the top. However, there are also some oxidic copper ores formed by weathering of sulfide ones (ores). Well known are cuprite, $Cu_2O$, and malachite, $Cu_2(OH)_2CO_3$. Occasionally, copper has been found in elementary form. Copper has two isotopes; namely, $^{63}Cu$ (69.1%) and $^{65}Cu$ (30.9%).

### Preparation

More than 80% of the processed copper ores have a copper content of less than 2%. Therefore, the ore must first be concentrated by floatation. Sulfide ores are then roasted with sulfur being partially converted into $SO_2$. At the same time, ingredients such as As, Sb, and Se are removed. The remainder mixture of mainly oxides and sulfides of Fe and Cu is molten under addition of $SiO_2$ and air. Iron sulfide is thus transferred into iron oxide, which is finally separated as slag. Copper sulfide is more resistant to air than FeS:

$$FeS + \frac{3}{2}O_2 + SiO_2 \rightarrow FeSiO_3 + SO_2$$

https://doi.org/10.1515/9783111574349-010

Now part of the copper sulfide is also converted into the oxide. The latter reacts with excess CuS under exclusion of air, yielding crude copper with a copper content of 98–99%:

$$Cu_2S + \frac{3}{2}O_2 \rightarrow Cu_2O + SO_2$$

$$Cu_2S + 2Cu_2O \rightarrow 6Cu + SO_2$$

Further refining is accomplished in two different ways. Dry refining starts with molten copper being treated with wood. Copper oxides are hereby reduced to the metal (99.95% copper content). About 80% of the crude copper is upgraded and purified electrolytically. In doing so, copper is made to operate as anode in an aqueous solution of $CuSO_4$ and sulfuric acid. At a voltage of 0.2–0.3 V, the anodically dissolved copper is separated as pure copper at the cathode. All the less noble impurities remain is solution, whereas the nobler metal settles as anode slime at the bottom of electrolysis cell. Due to the high cost of electric power consumed by electrolysis, it is only the noble metals in the anode mud which make the process economical one.

As the copper prices are rapidly increasing, the upgrading of poor copper ores becomes more and more interesting. Thus, extraction processes (hydrometallurgy) are about to gain industrial significance nowadays.

## Physical properties

Copper is ductile and relatively hard. It exhibits the highest thermal and electrical conductivity next to silver. Impurities decrease the conductivity but in turn increase the hardness of the metal. Volatile copper compounds, for example, the halides cause the green flame (Beilstein test for halides in organic compounds using a copper wire). The metal itself has a reddish color.

Table 1.7: Some physical properties of copper.

| Properties | $Z$ | Electronic configuration | Atomic weight | Density | m.p. | b.p | $\chi$ | Atomic radius |
|---|---|---|---|---|---|---|---|---|
| **Expression** | 29 | $[Ar]3d^{10}5s^1$ | 63.54 | 8.96 g/cm$^3$ | 1,085 °C | 2,570 °C | 1.90 | 128 pm |

## Chemical properties

Copper is relatively resistant to corrosion but attacked by air, giving a thin layer of oxide. In $CO_2$- or $SO_2$-containing wet atmosphere, green hydrocarbonates and hydroxosulfates are formed. At red heat, copper reacts with $O_2$ to give CuO, which is again decomposed to $Cu_2O$ at higher temperatures. Copper and sulfur form $Cu_2S$ or other nonstoichiometric sulfides. With halogens, the corresponding copper(II) halides are obtained. As copper is nobler than hydrogen, it cannot be dissolved by nonoxidizing acids. However, if oxygen is present, even these acids attack copper. Nitric and hot concentrated sulfuric acid dissolve copper under evolution of NO and $SO_2$, respectively:

$$3Cu + 2HNO_3 + 6H^+ \longrightarrow 3Cu + 2NO + 4H_2O$$

Copper has some physiological significance as a trace element. It is the constituent of certain oxidizing enzymes. Unlike other heavy metals, copper is only slightly toxic to human beings but causes vomiting after consumption of several hundred milligrams.

### Application

At present, copper is the most frequently used metal after iron and aluminum. Due to its excellent electrical conductivity, it is essential for the electroindustry but nowadays it is being gradually displaced by aluminum. Its high thermal conductivity and resistance to corrosion make it valuable for containers, heat engineering purposes, and so on. Finely dispersed copper, prepared by reduction of CuO with hydrogen, may be used for catalytic purposes (e.g., dehydrogenation of aldehydes) and for removal of oxygen traces from nitrogen.

### Copper(I) compounds

In general, copper(II) seems to be more stable than copper(I). However, the stability depends greatly on the reaction conditions (solubility, complex formation, etc.). The following equilibrium has an equilibrium constant of $10^6$ in aqueous solution:

$$2Cu^{2+}(aq) \rightleftharpoons Cu^{2+}(aq) + Cu(s)$$

The formation of sparingly soluble cuprous compounds or very stable complexes of Cu(I) may, however, shift it to left-hand side. $Cu^+$ shows a tendency for covalent bonding and, therefore, reacts preferably with polarizable anions or ligands. Copper(I) favors low coordination numbers in its compounds. Tetrahedral and linear ligand arrangements are commonly encountered. This is in contrast to $Cu^{2+}$, which prefers ionic bonding as well as square planar and distorted octahedral structures.

### Halides

Copper(I) has a $d^{10}$ configuration and its compounds are usually white or pale yellow. Cu(I) halides may be synthesized either by synproportionation of Cu(II) compounds and elementary copper, or by reduction of Cu(II) halides with sulfite or other suitable reducing agents.

$Cu_2O$ is the most stable copper oxide at high temperatures, and is generated from CuO above 1,200 °C. The sparingly soluble compound which is semiconductor can also be obtained by reduction of alkaline Cu(II) solutions. A modification of this reaction is used as a very sensitive test for organic reducing agents (aldehydes, aldehyde-containing sugars, etc.). Fehling's solution, which is recommended for this test, contains $CuSO_4$, NaOH, and sodium ammonium tartrate as a strong complexing ligand. The latter is necessary in order to avoid precipitation of $Cu(H_2O)$. In this way, it is possible to detect 1 µg of sugar (e.g., in human blood):

$$4Cu^{2+} + HCHO\ 4OH^- \rightarrow 4Cu^+ + CO_2 + 5H_2O$$

The unstable, yellow CuOH is immediately converted into red $Cu_2O$. $Cu_2S$ is a black solid formed from the elements at higher temperatures.

### Other copper(I) compounds and complexes

$Cu^{2+}$ reacts with cyanide in a similar manner as with iodide. Reduction takes place under precipitation of yellow CuCN and also evolution of gaseous dicyane $(CN)_2$. The white CuCN can also be obtained:

$$2Cu(CN)_2 \rightarrow 2CuCN + (CN)_2$$

## 9.4 Chemical and apparatus

| Chemicals | Apparatus |
|---|---|
| Copper metal and grain of zinc | Test tube |
| $CuSO_4$ | Bunsen burner |
| Iron nail | |
| $AgNO_3$ solution and copper wire | |
| 2N, HCl | |
| 2N, $H_2SO_4$ | |
| 2N, $HNO_3$ | |
| Copper(II) oxide | |
| NaOH and $NaCO_3$ solution | |
| Aldehyde solution and KI solution | |

## 9.5 Procedure

### 1.1 Preparation of copper from copper(II) oxide

On a sheet of paper thoroughly mix two to three micro-spatulas of Cu and an equal amount of powdered charcoal. Transfer the mixture to a test tube and heat strongly for 5 min. Observe the reduction to brown, metallic copper:

$$2CuO + C \rightarrow Cu + CO_2$$

If it proves to be difficult to clean the test tube, use some drops of concentrated nitric acid. This acid dissolves the remaining copper completely.

### 1.2 Preparation of copper by displacement from copper(II) salt solution with zinc and iron

Take five to six drops of $CuSO_4$ in a test tube and add grain of zinc. Shake gently and observe the appearance of a red color deposit on the surface of zinc. Repeat the experiment using iron nail in place of zinc:

$$Cu^{2+} + Zn \rightarrow Cu + Zn^{2+}$$

### 1.3 Reducing properties of metallic copper: displacement of silver from Ag(I) solution by copper

Add five to six drops of $AgNO_3$ solution to a test tube and dip end of a cleaned copper wire. You will observe grayish spongy deposition of white substance together with some lustrous crystal of silver on the surface of the copper wire:

$$Cu + 2Ag^{+} \rightarrow Cu^{2+} + 2Ag$$

### 1.4 Action of dilute and concentrated acids on copper

Add a small piece of copper to each of three test tubes and then five to six drops of 2N, HCl to one test tube, 2N, $H_2SO_4$ to the second test tube, and 2N, $HNO_3$ to the third one. The color of the solution in the third test tube changes to green, which indicates the formation of $Cu^{2+}$ and brown gas involved implies $NO_2$ formed from NO and air.

**Note:** Only 2N, $HNO_3$ dissolves copper:

$$Cu + HNO_3 + 6H^{+} \rightarrow 3Cu^{2+} + NO + 4H_2O$$

Repeat this experiment with concentrate HCl, $H_2SO_4$, and $HNO_3$. Note that Cu dissolves both in concentrated $H_2SO_4$ and $HNO_3$:

$$Cu + H_2SO_4 \longrightarrow CuSO_4 + SO_2 + 2H_2O$$

## 2 Copper(II) compounds

### 2.1 Reaction of copper(II) oxide with dilute acids

Add two to three micro-spatula of copper(II) oxide to each of two test tubes, then add five to six drops of 2N, HCl to one test tube and equal amounts of 2N, $H_2SO_4$ to the other test tube. What can be concluded from the color of the resulting solution? Write the equation.

### 2.2 Preparation of copper(II) hydroxide and its reaction with acids and bases, and thermal decomposition

Add five to six drops of $CuSO_4$ solution to three test tubes and then three to four drops of NaOH to all three test tubes. Blue $Cu(OH)_2$ precipitates will be observed.

Then heat one of the test tubes until black color appears. Take the remaining two tubes, add 2N, NaOH to one test tube and equal amount of 2N, $H_2SO_4$ to the other.

The acid completely dissolves $Cu(OH)_2$ to give blue solution of $Cu^{2+}$ while the base dissolves only partially by forming dark blue tetrahydroxocuprate.

Reactions:

$$Cu^{2+} + 2OH^- \longrightarrow Cu(OH)_2 \xrightarrow{\text{heat}} CuO + H_2O$$

$$Cu(OH)_2 + 2H^+ \longrightarrow Cu^{2+} + 2\,H_2O$$

$$Cu(OH)_2 + 2OH^- \longrightarrow [Cu(OH)_4]^{2-}$$

### 2.3 Hydrolysis of copper(II) salts

Place two drops of $CuSO_4$ solution and the same volume of $CuCl_2$ on a blue litmus paper. The litmus paper turns to red:

$$[Cu(H_2O)_6]^{2+} + H_2O \rightarrow [Cu(H_2O)_5(OH)] + H_3O^+$$

### 2.4 Sparingly soluble Cu(II) salts: CuS

Add two to three drops of $CuSO_4$ solution to a test tube and the same amount of fresh $H_2S$ water. Black CuS mixed with $Cu_2S$ is precipitated:

$$Cu^{2+} + S_2^{\,-} \rightarrow CuS$$

### 2.5 Preparation of copper(II)carbonate ($CuCO_3$)

Add two to three drops of $CuSO_4$ solution to a test tube and the same amount of soda $NaCO_3$ solution. Observe the precipitation of blue basic copper carbonate, $Cu(OH)_2{\cdot}CaCO_3$:

$$Cu^{2+} + CO_3^{\,2-} + H_2O \rightarrow Cu(OH)_2 \cdot CuCO_3$$

### 2.6 Complex formation of copper(II): reaction with ammonia

Place two to three drops of $CuSO_4$ solution into a test tube and add dropwise ammonia solution. At first, blue $Cu(OH)_2$ precipitates but dissolves in excess ammonia under the formation of deep blue tetrammine complex, $[Cu(NH_3)_4]^{2+}$. This reaction is overly sensitive and specific for Cu(II):

$$Cu^{2+} + 2OH^- \rightarrow Cu(OH)_2$$

$$Cu(OH)_2 + 4NH_3 \rightarrow [Cu(NH_3)_4]^{2+} + 2OH^-$$

## 3 Copper(I) compounds

### 3.1 Preparation of Cu(I) hydroxide and its thermal decomposition

Take two to three drops of $CuSO_4$ solution and add five to six drops of 10% aldehyde solution. Heat the mixture to boiling and treat with four to five drops of 2N, NaOH.

**Note:** The precipitation of yellow CuOH. Continue heating until the color of the solid turns to red, which indicates the presence of $Cu_2O$:

$$Cu(OH)_2 + 4HCOH \rightarrow CuOH + CO_2 + 3H_2O$$

**Note:** Similarly CuOH is thermally unstable and decomposes to $Cu_2O + H_2O$:

$$CuOH \rightarrow CU_2O + H_2O$$

### 3.2 Preparation of cuprous iodide, CuI

Add two drops of $CuSO_4$ solution and the same amount of KI solution to a test tube. White CuI precipitates and the mixture turns to yellow to brown due to the formation of iodine (as $I_3^-$). Use starch solution to prove the presence of iodine; touch the sample with a glass rod and dip it into another test tube containing some drops of starch solution. The latter will turn blue.

To observe the white color of CuI, add several drops of $Na_2SO_3$ solution. $Na_2SO_3$ removes iodine and white color appears, hence, keep it for further experiments:

$$Cu^{2+} + 2I^- \rightarrow CuI + \frac{1}{2}I_2 \qquad I_2 + I^- \rightarrow I_3^-$$

$$I_2 + SO_3^{2-} + H_2O \rightarrow 2I^- + SO_4^{2-} + 2H^+$$

## 9.6 Observation

Color change
Evolution of gas
Sound of reaction when reaction takes place

## 9.7 Results and discussion

## 9.8 Review questions

1. Indicate the position of copper in the periodic table and work out its electronic configuration.
2. The electronic configuration of group IB elements is quite similar to that of alkali elements ($ns^1$). Compare the properties of both groups of elements and find the similarities in their behavior.
3. Why copper and silver are poorer reducing agents than the alkali metals?
4. Which are the most common oxidation states of copper in its compounds?
5. How can you distinguish between Cu(I) and Cu(II) compounds by making uses of their magnetic properties?
6. What reaction, if any, takes place on treating copper with:
   a) dilute sulfuric acid and b) hot conc. $H_2SO_4$?
7. Dry copper chloride is yellow-brown. A conc. aqueous solution of the halide has green color. On adding conc. $H_2SO_4$ or HCl, the color changes to yellow-brown. Dilution with water creates a blue color. Explain these color changes.
8. Why is it possible to use anhydrous $CuSO_4$ as an indicator for traces of moisture?

# Experiment 10

## 10.1 Chemistry of zinc

## 10.2 Objective

Students should be able to:

1. understand how you will prepare metallic zinc from ZnS and $ZnCO_3$,
2. write the equation for the reaction of zinc with
   a. hydrochloric acid,
   b. dilute and concentrated sulfuric acid,
   c. dilute and concentrated nitric acid, and
   d. alkali hydroxide, and
3. characterize the behavior of $Zn(OH)_2$ with respect to its solubility in acids and bases (NaOH and $NH_3$).

## 10.3 Theory

### Occurrence

Zinc is widespread in nature with a portion of 120 g/t in the Earth's crust (similar to copper). Zinc occurs most frequently in the form of sulfides (sphalerite, ZnS), carbonates (galmei, $ZnCO_3$), and also as silicates and oxides. Most of zinc ores contain about 0.3% Cd. Zinc is known with isotopes $^{64}Zn$ (48.9%), $^{66}Zn$ (27.8%), $^{67}Zn$ (4.1%), $^{68}Zn$ (18.6%), and $^{70}Zn$ (0.6%).

### Preparation

As brass alloy (Cu/Zn), zinc has been known since antiquity. The preparation of the pure metal, however, proved to be difficult because of its considerable volatility. Zinc ores are normally concentrated by floatation followed by roasting to give the oxide:

$$ZnS + \frac{3}{2}\,O_2 \rightarrow ZnO + SO_2$$

$$ZnCO_3 \rightarrow ZnO + CO_2$$

The preparation of the metal is brought about thermally or electrolytically:

$$ZnO + C \rightarrow Zn + CO$$

$$ZnO + CO \rightarrow Zn + CO_2$$

https://doi.org/10.1515/9783111574349-011

The electrolytic method starts with ZnO being dissolved in sulfuric acid followed by electrolysis to give the metal.

## 10.4 Chemical and apparatus

| Chemicals | Apparatus |
|---|---|
| - Zinc dust | Test tube |
| - Solution of zinc sulfate or chloride. | Bunsen burner |
| - 2N, $H_2SO_4$ | |
| - Concentrated $H_2SO_4$ | |
| - 2N, NaOH | |
| - 0.5 N $KNO_2$ | |
| - 2N, $NH_3$ | |
| - Acetic acid and sodium acetate | |

## 10.5 Procedure

### 1.1 Dissolution of zinc in $H_2SO_4$

Add a micro-spatula of zinc dust to four to five drops of 2N, $H_2SO_4$ in a test tube and heat the mixture gently in a water bath until you note the evolution of hydrogen gas, which leads to the formation of colorless $Zn^{2+}$ solution:

$$Zn + 2H^+ \rightarrow Zn^{2+} + H_2$$

Repeat the above experiment using concentrated $H_2SO_4$.

**Note:** Zinc does not dissolve readily in concentrated $H_2SO_4$ and only reacts slowly on heating to form $ZnSO_4$ and $SO_2$:

$$Zn + 2H_2SO_4 \rightarrow ZnSO_4 + SO_2 + H_2$$

### 1.2 Dissolution of zinc in alkali hydroxide

Add a micro-spatula of zinc dust to four to five drops of 2N, NaOH in a test tube and heat the mixture gently in a water bath. Zinc dissolves by giving colorless solution, $[Zn(OH)_3]^-$, liberating hydrogen gas:

$$Zn + 2H_2O + 2OH^- \rightarrow [Zn(OH)_3]^- + H_2$$

### 1.3 Reducing property of zinc: reduction of nitrite to ammonia

In a porcelain crucible containing a micro-spatula of zinc dust, add four drops of 0.5 N $KNO_2$ followed by four to five drops of concentrated NaOH. Heat the mixture gently to boil on asbestos wire gauze. You will feel the smell of pungent odor due to the formation of ammonia. Check using the red litmus paper:

$$Zn + NO_2^- + 5H_2O + 2OH^- \rightarrow [Zn(OH)_3]^- + NH_3$$

### 1.4 Reaction of Zn(II) with ammonia: formation of tetrammine zinc(II)

Add three drops of Zn(II) solution in a test tube, and then 2N $NH_3$ in dropwise manner. White $Zn(OH)_2$ precipitates in the first instance but dissolves in excess amount of $NH_3$:

$$Zn(OH)_2 + NH_3 \rightarrow [Zn(NH_3)_4]^{2-} + 2OH^-$$

### 1.5 Sparingly soluble zinc salts: zinc carbonate

Add equal amount of sodium carbonate to three to four drops of zinc salt solution in a porcelain crucible. White zinc carbonate precipitates and heat gently; carbonate and hydroxides of Zn decompose well. Write the equation.

### 1.6 Preparation of zinc sulfide

Add a drop of acetic acid and sodium acetate into a test tube containing three to four drops of zinc salt solution and treat with a few drops of freshly prepared $H_2S$ solution. Note the white precipitate of zinc sulfide:

$$Zn^{2+} + H_2S \rightarrow ZnS + 2H^+$$

Repeat the above experiment using two to three drops of $H_2SO_4$ in place of acetic acid and sodium acetate. No precipitate is formed, but what is the reason behind? Explain.

## 10.6 Observation

Color change
Evolution of gas
Sound of reaction when reaction takes place

# Experiment 11

## 11.1 Chemistry of cadmium

## 11.2 Objective

Students should be able to:
1. observe the reaction of cadmium(II) with aqueous sodium hydroxide,
2. characterize the behavior of $Cd(OH)_2$, and
3. understand the precipitation of cadmium sulfide.

## 11.3 Theory

### 11.3.1 The element

Cadmium is always associated with zinc. Zinc ores normally contain 0.2–0.4% cadmium. The element is used for protective layers on other metals. Its production appears to be quite complex and starts with the oxide or the sulfate. Cadmium oxide can be reduced by carbon, and $CdSO_4$ is processed electrolytically. The latter is soluble in dilute nitric acid but only slowly dissolved by HCl and $H_2SO_4$.

Additionally cadmium has eight isotopes: $^{106}Cd$ (1.2%), $^{108}Cd$ (0.9%), $^{110}Cd$ (12.4%), $^{111}Cd$ (12.7%), $^{112}Cd$ (12.4%), $^{113}Cd$ (12.3%), $^{114}Cd$ (28.9%), and $^{116}Cd$ (7.6%).

### 11.3.2 Cadmium compounds

The only stable oxidation state of cadmium in its compounds under normal condition is +II. In all, its reaction cadmium is quite similar to zinc.

Cadmium hydroxide is obtained as white precipitate on adding NaOH or ammonia to an aqueous Cd(II) salt solution. In contrast to zinc hydroxide, $Cd(OH)_2$ is not soluble in an excess of alkali hydroxide. It dissolves, however, in aqueous ammonia under formation of the ammine complex.

The oxide, CdO, can be prepared from $Cd(OH)_2$ by heating above 150 °C. Its brown color is a result of lattice defects.

Cyanide and $Cd^{2+}$ give a white precipitate of $Cd(CN)_2$, which reacts with excess cyanide yielding the soluble complex anion, $[Cd(CN)_4]^{2-}$.

Brown yellow CdS is precipitated from slightly acidic (mineral acid) aqueous solutions of Cd(II) salts with $H_2S$.

https://doi.org/10.1515/9783111574349-012

Notice that ZnS has a slightly higher solubility and its precipitation requires acetic acid medium. CdS seems to be rather inert to complexing agents and is not attacked by cyanide (difference to CuS).

## 11.4 Chemicals and apparatus

| Chemicals | Apparatus |
|---|---|
| – A solution of cadmium sulfate or chloride | Test tube |
| – NaOH | Bunsen burner |
| – 2N ammonia | |
| – 2N sulfuric | |
| – $H_2S$ | |

## 11.5 Procedure

### 1 Reaction of cadmium(II) with aqueous sodium hydroxide

Add three to four drops of Cd(II) salt solution to a test tube and then add NaOH in a dropwise manner. Observe the white precipitate of $Cd(OH)_2$. In contrast to Zn, Cd(OH)2 is not soluble in excess NaOH:

$$Cd^{2+} + 2OH^- \rightarrow Cd(OH)_2$$

Find an explanation for the different behavior of $Zn(OH)_2$ and $Cd(OH)_2$.

### 2 Reaction of cadmium(II) with ammonia: formation of a hexammine complex

Repeat the previous experiment but use 2N ammonia instead of aqueous alkali hydroxide. Initially, white precipitate of $Cd(OH)_2$ appears, which is, however, dissolved in excess ammonia. The colorless hexammine complex $[Cd(NH_3)_6]^{2+}$ is formed. Zinc exhibits the same behavior. Write down the equations.

### 3 Preparation of basic cadmium carbonate

Add three to four drops of Cd salt solution into a porcelain crucible followed by an equal volume of sodium carbonate solution. Observe the white precipitate of basic cadmium carbonate. The compound has variable compositions. After evaporating the solvent and heating residue, a color change from white to brown will occur. This is attributed to lattice defects in CdO, which has been formed by thermal decomposition of the basic cadmium carbonate.

Write down the equations.

### 4 Precipitation of cadmium sulfide

Add three to four drops of Cd(II) salt solution to the test tube, and then one drop of 2N sulfuric acid followed by two to three drops of freshly prepared H2S water. Observe the brownish yellow precipitate of CdS. The latter is soluble in semiconcentrated mineral acids.

## 11.6 Observation

- Color change
- Evolution of gas
- Sound of reaction when reaction takes place

## 11.7 Review problems

1. Describe the position of cadmium in the periodic table and find out its electronic configuration.
2. Which hydroxide is more basic, $Zn(OH)_2$, or $Cd(OH)_2$? Give a reason for your choice.
3. What happens to a solution containing $Zn^{2+}$ and $Cd^{2+}$ ions on adding excess NaOH or ammonia? Is it possible to distinguish between the two ions?

# Experiment 12

## 12.1 Chemistry of mercury

## 12.2 Objective

Students should be able to:
- understand the reaction of mercurous salts with sodium hydroxide and
- observe the reaction of mercury with hydrochloric acid.

## 12.3 Theory

### Occurrence

Mercury occurs only 0.5 g/t in the Earth's crust and, therefore, represents a rare element. Nevertheless, some relatively high deposits are known, which contain mercury in the form of red cinnabar, HgS. This is occasionally mixed with small droplets of mercury.

### Preparation

After flotative concentration of the mercury ore, the sulfide is roasted in a stream of air followed by condensation of the mercury vapor passing over:

$$HgS + O_2 \longrightarrow Hg + SO_2$$

The crude mercury is washed with dilute $HNO_3$, filtered, and distilled in vacuum.

### Physical properties

Mercury is a white-lustrous metal and liquid at room temperature. As metal, it proves to be comparatively volatile with a vapor pressure of 0.0016 mbar at 200 °C (for Cd only 10–10 mbar). Mercury and bismuth exhibit the highest specific resistance of all metals. Furthermore, Hg has a remarkably high surface tension compared to other liquids.

### Chemical properties

With respect to its chemical behavior, Hg differs considerably from its homologs Zn and Cd. As a noble one, it is resistant to $O_2$, water, $CO_2$, $SO_2$, HCl, HF, and ammonia at room temperature although it reacts with halogens and sulfur under these conditions. Nonoxidizing acids do not attack mercury, but $HNO_3$ and conc. sulfuric acid dissolve the metal. Due to its liquid state, mercury alloys with many other metals even at room temperature to form an amalgam. Very good precursors of amalgams are alkali and

https://doi.org/10.1515/9783111574349-013

alkaline earth metals, whereas 3d elements (except Mn and Cu) are nearly insoluble in liquid mercury. Mercury and all its soluble and volatile compounds are highly toxic as it reacts with human proteins to form Hg–S bond, which blocks the protein functions completely.

## 12.4 Chemicals and apparatus

| Chemicals | Apparatus |
|---|---|
| Aqueous solutions of $Hg_2(NO_3)_2$ and $Hg(NO_3)_2$ | Dropper |
| Ammonia solution | Porcelain crucible |
| Hydrochloric acid | Test tube |

## 12.5 Procedure

### 12.5.1 Displacement of mercury from its salts

Clean a copper plate with abrasive paper, place a drop of mercuric nitrate solution on it, and let it rest for 5 min. Rub the gray spot with filter paper to make it lustrous:

$$Hg^{2+} + Cu \rightarrow Hg + Cu^{2+}$$

### 12.5.2 Reaction of mercurous salts with sodium hydroxide

Take three drops of the mercurous salt solution in a test tube and add dropwise 2N, NaOH. The black precipitate consists of a mixture of Hg and HgO. It is soluble in nitric acid but insoluble in excess NaOH:

$$Hg_2{}^{2+} + 2OH^- \rightarrow Hg + HgO + H_2O$$

### 12.5.3 Reaction with hydrochloric acid

Add three drops of the mercurous nitrate solution to a test tube followed by a few drops of HCl. A white precipitate of $Hg_2Cl_2$ is obtained. The compound is not soluble in dilute mineral acids:

$$2Hg_2{}^{2+} + 2Cl^- \rightarrow Hg_2Cl_2$$

Addition of some drops of ammonia to thus

## 12.6 Review problems

1. Compile some of the characteristic group properties of Zn, Cd, and Hg. In which way does mercury differ from the other two elements?
2. How do you prepare metallic zinc from ZnS and $ZnCO_3$?
3. Write down the equation for the reaction of zinc with:
   a. hydrochloric acid,
   b. dilute and concentrated sulfuric acid,
   c. dilute and concentrated nitric acid, and
   d. alkali hydroxide.
4. Characterize the behavior of $Zn(OH)_2$ with respect to its solubility in acids and bases (NaOH, $NH_3$).
5. Zinc reduces nitrate and nitrite to ammonia in alkaline aqueous solution. Write down the equations.
6. Which solution, $H_2S$ or $(NH_4)_2S$, would more completely precipitate ZnS from a solution of its salt?
7. A solution of $ZnSO_4$ in water is acidic. Why?
8. Which salt, $ZnCl_2$ or $[Zn(NH_3)_4]Cl_2$, is hydrolyzed more strongly? Prove your answer.
9. Write down the various types of salts that can be formed from $Zn(OH)_2$.

# References

[1] Cotton, F. A. and Wilkinson, G. (1972). Advanced Inorganic Chemistry (3rd ed.). Inter science Publishers, a division of John Wiley and Sons, New York.

[2] Chambers, C. and Holliday, A. K. (1975). Modern Inorganic Chemistry (1st ed.). R. I. Acford Ltd., Industrial Estate, Chichester, Sussex.

[3] Housecroft, C. E. and Sharpe, A. G. (2005). Inorganic Chemistry (2nd ed.). Pearson Education Limited, England.

https://doi.org/10.1515/9783111574349-014

# Index



https://doi.org/10.1515/9783111574349-015

www.ingramcontent.com/pod-product-compliance
Lightning Source LLC
Chambersburg PA
CBHW081548220326
41598CB00036B/6602
* 9 7 8 3 1 1 1 5 7 3 8 4 7 *